CONQUEST IN CYBERSPACE

The global Internet has served primarily as an arena for peaceful commerce. Some analysts have become concerned that cyberspace could be used as a potential domain of warfare, however. Martin C. Libicki argues that the possibilities of hostile conquest are less threatening than these analysts suppose. It is in fact difficult to take control of other people's information systems, corrupt their data, and shut those systems down. Conversely, there is considerable untapped potential to influence other people's use of cyberspace, as computer systems are employed and linked in new ways over time.

The author explores both the potential for and limitations to information warfare, including its use in weapons systems and in command-and-control operations as well as in the generation of "noise." He also investigates how far "friendly conquest" in cyberspace extends, such as the power to persuade users to adopt new points of view. Libicki observes that friendly conquests can in some instances make hostile conquests easier or at least prompt distrust among network partners. He discusses the role of public policy in managing the conquest and defense of cyberspace and shows how cyberspace is becoming more ubiquitous and complex.

Martin C. Libicki, a senior policy analyst at the RAND Corporation since 1998, works on the relationship between information technology and national security. He has written numerous monographs on the subject, notably *What Is Information Warfare*, *The Mesh and the Net: Speculations on Armed Conflict in a Time of Free Silicon*, and *Who Runs What in the Global Information Grid*. Dr. Libicki is also the editor of the RAND textbook *New Challenges: New Tools for Defense Decisionmaking*. His most recent assignments at RAND have been to generate novel information system capabilities for counterinsurgency and to develop a post-9/11 information technology strategy for the U.S. Department of Justice and the Defense Advanced Research Projects Agency's (DARPA) Terrorist Information Awareness program; to conduct an information security analysis for the FBI; to investigate targeting strategies of al Qaeda; and to assess the CIA's research and development venture, In-Q-Tel. He previously worked at the National Defense University, was on the Navy Staff as program sponsor for industrial preparedness, and was a policy analyst for the Government Accountability Office's Energy and Minerals Division. Dr. Libicki received his Ph.D. from the University of California at Berkeley in 1978.

Conquest in Cyberspace

National Security and Information Warfare

MARTIN C. LIBICKI

The RAND Corporation

CAMBRIDGE
UNIVERSITY PRESS

CAMBRIDGE UNIVERSITY PRESS
Cambridge, New York, Melbourne, Madrid, Cape Town, Singapore,
São Paulo, Delhi, Dubai, Tokyo, Mexico City

Cambridge University Press
32 Avenue of the Americas, New York, NY 10013-2473, USA

www.cambridge.org
Information on this title: www.cambridge.org/9780521692144

First published 2007

A catalog record for this publication is available from the British Library

Library of Congress Cataloging in Publication data
Libicki, Martin C.
Conquest in cyberspace : national security and information warfare / Martin C. Libicki ;
RAND Corporation.
p.   cm.
Includes bibliographical references and index.
ISBN-13: 978-0-521-87160-0 (hardback)
ISBN-13: 978-0-521-69214-4 (pbk.)
1. Information warfare.   2. National security.   3. Cyberterrorism.
4. Computer networks – Security measures.   I. Rand Corporation   II. Title.
U163.L534   2007
355.3´43 – dc22   2006030973

ISBN 978-0-521-87160-0 Hardback
ISBN 978-0-521-69214-4 Paperback

Contents

List of Figures

Acknowledgments

Perhaps the greatest joy in working for the RAND Corporation is the opportunity to work with interesting, intelligent, and inquisitive colleagues. When collaboration works, and it often does, it is far easier to determine with whose pen thoughts were rendered in English than to discern from whose mind such thoughts came. Three such colleagues merit note here, not least because this manuscript would never have been written without them.

James Mulvenon suggested that we work together on a project to define exactly what information warfare (IW) is. The trick in such endeavors is to hew to the art of the technically possible, without, at the same time, basing theory on the evanescent characteristics of today's information technology. Chapters 2, 3, and part of 11 arose from our joint work. We also worked together on another project that looked at what light a theory of command and control could shed on information warfare. Chapter 5 reflects that work.

David Frelinger arranged for us to think systematically about what an information warfare attack on an integrated air defense system (IADS) would look like. The question was prompted by inquiries over whether one could quantify the effects of information warfare on an IADS with as much confidence as one could for the effects of electronic or physical warfare. Short answer: no. Chapter 4, which deals broadly with information warfare against critical systems, grew out of the initial efforts to explain why not.

Laurent Murawiec led me into other chapters of the manuscript through a joint project that looked for a theory of command and control of the sort that the Pentagon's Office of Net Assessment could use. Under his prompting, I developed the material that now constitutes parts of Chapters 1, 6, 8, and 10.

Big thanks are also due to those who reviewed and commented on the manuscript in its various incarnations: Paul Davis, Robert Klitgaard, Shari Lawrence Pfleeger, Charles Wolf, all at RAND, and Professor Anthony Oettinger of Harvard. In addition, Judy Lewis and Lisa Sheldone at RAND have been an invaluable source of assistance and support in making the review and publishing process work well.

Introduction

Despite its roots in the U.S. Department of Defense (DoD), the global Internet has primarily, although not exclusively, been an avenue and arena of peaceful commerce. With every year, an increasing percentage of the world's economy has migrated from physical media, or older electronic media such as telephones and telegraphs, to the public Internet and to private or semipublic internets. Systems that were once inaccessible to persons off-premises, such as power plant controls, are now theoretically accessible to anyone around the world. Other hitherto self-contained networks, such as those that transferred money, are now commingled with the larger, more public networks such as the Internet or the international phone system.

Indeed, its very success is what has turned the Internet into a potential venue of warfare. It is not only that defense systems of advanced militaries are being knit into more powerful systems of systems – thereby becoming the militaries' new center of gravity. The real impetus is that the more cyberspace is critical to a nation's economy and defense, the more attractive to enemies is the prospect of crippling either or both via attacks on or through it. Hackers can and do attack information systems through cyberspace. They can attack the cyberspace itself through operations against the networks that provide the basis for this new medium. Defenders thus must keep these hackers out of their systems. If hackers get in, they could wreak great damage. At a minimum they might steal information. Worse, they can make systems go haywire. Worst, they could inject phony information into systems to distort what users think they absorb when they deal with systems. Hackers might take over any machine (such as a pump) controlled by a networked computer system and use it according to their ends and not those of its owners.

None of this requires mass, just guile. For that reason, attacks in cyberspace do not need the same government backing as attacks in older media do. Any group, or even individual, can play – even, perhaps especially, terrorists. Prior to 9/11, in fact, it was difficult to conceive of a strategic attack on the U.S. homeland by nonstate actors *except* through the medium of cyberspace. Such would be a bloodless attack from afar that left no traces but could cause the systems we rely on to crash mysteriously. The President's Commission on Critical Infrastructure Protection argued in 1996 that the capability to launch such an attack did not yet exist – but given five years (that is, by 2001), it very well might.

Perhaps needless to add, although advanced nations have more at stake in cyberspace than developing nations do, the latter are increasingly being drawn into its domain. Thus, they too are vulnerable to attacks from what are, in general, the larger and more sophisticated cohorts of hackers from the first world.

By such means, cyberspace has joined air and outer space as a new medium of conflict.[1] Granted, evidence that it has become a *significant* medium of conflict is sparse. This may be because the last three wars in which cyberspace could have played a role – Kosovo, Afghanistan, and Iraq, respectively – were against countries with minimal presence in cyberspace. They had little that the United States could attack, or at least attack more efficiently than conventional means already permitted it to do. So far, other countries have lacked the sophistication and will to do much damage to the U.S. use of cyberspace. But since participation

---

[1] The 2001 Department of Defense Quadrennial Defense Review Report listed four "Key Military–Technical Trends." The third was "Emergence of new arenas of military competition":

Technological advances create the potential that competitions will develop in space and cyber space. Space and information operations have become the backbone of networked, highly distributed commercial civilian and military capabilities. This opens up the possibility that space control – the exploitation of space and the denial of the use of space to adversaries – will become a key objective in future military competition. Similarly, states will likely develop offensive information operations and be compelled to devote resources to protecting critical information infrastructure from disruption, either physically or through cyber space (p. 7).

in and dependence on cyberspace is growing, the odds of consequential conflict, and thus hostile conquest, must certainly be rising.

Lost in this clamor about the threat from hackers is another route to conquest in cyberspace, not through disruption and destruction but through seduction leading to asymmetric dependence. The seducer, for instance, could have an information system attractive enough to entice other individuals or institutions to interact with it by, for instance, exchanging information or being granted access. This exchange would be considered valuable; the value would be worth keeping. Over time, one side, typically the dominant system owner, would enjoy more discretion and influence over the relationship, with the other side becoming increasingly dependent. Sometimes the victim has cause to regret entering the relationship; sometimes all the victim regrets is not receiving its fair share of the joint benefits. But if the "friendly" conquest is successful, the conqueror is clearly even better off.

The central contention of this work is that the possibilities of hostile conquest may be less consequential than meets the eye while the possibilities of friendly conquest ought to be better appreciated. The current obsession with hostile conquest fosters a tilt toward closed systems, at least among those who have powerful systems to begin with. Those with the most attractive systems – in terms of information, knowledge, services, and reach – have an inherent advantage whose benefits they might deny themselves by concentrating on the threat to themselves. This is particularly so for the national and homeland security community (including law enforcement, homeland defense, and infrastructure). By taking a more open approach to cyberspace, they may extend their influence and the influence of their values more certainly than they would by taking a closed approach.

In a sense, this argument echoes the distinction made by Joseph Nye between a nation's hard power and its soft power.[2] Hard power is embodied in military force, soft power in its culture. Hard power, like hostile conquest in cyberspace, ultimately entails one nation doing to another what the other would prefer it not do. It is involuntary. Soft power, like

---

[2] Joseph Nye, *Bound to Lead: The Changing Nature of American Power*, New York (Basic Books), 1990.

friendly conquest in cyberspace, describes the process of enticement. It is voluntary, at least at first. In the case of soft power, the elites of the affected country may find themselves unable to roll back the tide of imported cultural and economic mores without facing resistance and revolt. But rarely can one nation control or manipulate the instruments of soft power to create such a dependency; more often, it works independent of national strategy. With friendly conquest in cyberspace, however, the seducer retains part of the leverage precisely because the controls over the seductive system are not relinquished.

Hence the choices, many of them public choices. Hence, too, the orientation of this work, one to be understood in its policy and management rather than technical context. It is aimed at educated individuals who are interested in public policy. Admittedly, issues of cyberspace can become quite technical, and so the text tries to clarify some key concepts. Cyberspace issues are not unique in that regard. It can be hard to understand, say, the pros and cons of strategic ballistic missile defense without some understanding of physics. Nevertheless, arguments about strategic defense are not entirely technical ones. Similarly, arguments about the proper use and exploitation of cyberspace are not entirely technical. Readers who happen to be information security experts may appreciate reading this or that point of view; they are unlikely to add much to their technical knowledge of their craft by reading this.

1.1 What Does Conquest Mean in Cyberspace?

This work is entitled not "The Conquest *of* Cyberspace" but "Conquest *in* Cyberspace" for a reason. To emphasize the "of" is to suggest that there is, in fact, *a* cyberspace that exists in the same sense that the oceans do. It has distinct parameters and perimeters, and one can define conquest within this space. This leaves the only interesting question one of determining who has, in fact, taken possession of what part of cyberspace and how they accomplished such feats. Emphasizing "in," by contrast, reflects the fact that while something akin to conquest can be defined for cyberspace, cyberspace itself cannot be conquered in any conventional sense.

To understand why, it helps to understand what cyberspace itself means. Ironically, that process is best begun by discussing what cyberspace does *not* mean – or at least does not mean yet.

The term "cyberspace" was coined in William Gibson's 1984 classic, *Neuromancer*. The concept was further described in compelling detail in Neil Stephenson's 1989 *Snow Crash*. Both portrayed it as an alternative universe that people could participate in ("jack into" pace Gibson). It may be seen, particularly in some movies, as being just on the other side of the twenty-first century's version of Alice's looking-glass. Cyberspace, so defined, may be evoked through a text-only medium such as a chat room, but it can also be evoked more tangibly by a virtual reality simulation in which what one sees, hears, and, to some extent, feels is all synthesized on the spot. Computer power and fat networks make this illusion easier to generate with every passing year.

This often attractive concept should not lead one to imagine cyberspace as being *the* parallel universe – as if a mapping of this reality into another dimension. Four tenets suggest why cyberspace should be understood on its own merits.

First, cyberspace is a replicable construct. Being replicable, it exists in multiple locations at once. Because it is replicable, it is also reparable.

By contrast, only confusion can follow the unconscious assumption that there is *one* cyberspace in the sense that there is, say, one outer space. The existence of a single something called outer space derives from the simple fact that there is a planet earth and that every point on or above the planet has a unique location relative to it. This uniqueness is firmly rooted in physical law. The planet, for instance, has only one geosynchronous belt, and locations[3] in it are carefully allocated for every satellite (of a given broadcast frequency). There is also one spectrum, uses of which are governed by international conventions such as the World Radio Conference. From a military perspective, one nation's fleets of hunter-killer satellites can keep another country from establishing its own constellation. Control in space, can, in theory, be exclusive.

Cyberspace, by contrast, is built, not born. Every system and every network can hold its own cyberspace – indeed, it can hold a limitless number of quasi-independent spaces. Cyberspace can appear in multiple,

---

[3] Satellites in geosynchronous orbit appear to linger above a single point on the equator. Satellites in such orbits have to be separated from each other by a certain arc length if they broadcast in the same frequencies. As such, there are a finite number of such orbits and each is assigned on a global basis.

almost infinite, manifestations and forms. Even shared spaces can be indefinitely replicated. This is apparent, for instance, in multiplayer games (such as the *Sims Online* or *Rise of Nations*). Since the number of players in any one game is small, there have to be many of them to accommodate everyone who wants to play.

Not only is cyberspace a construct, but the rules of cyberspace are largely constructs – there is little hard-and-fast physics of the sort that dictates what can and cannot be done in, say, outer space. What can and cannot be done in cyberspace need not follow the laws of physics or the laws of man – although violating the latter may have real-world repercussions. There is no inherent "there" there except as mutually accepted.[4] Even larger games (massive multiplayer online role-playing games [MMOPRGs]) such as *EverQuest* or the popular South Korean multiplayer game *Lineage*, although more unified, exist because they have been constructed for that purpose, often a commercial one. Admittedly, some people are so hooked by these games that they actually pay real money to acquire virtual goods, useful only online.[5] Yet, they are not inherently fixed; should something more attractive come along, they could be easily abandoned. Not so for outer space or the oceans; they will always be there.

Second, to exist in cyberspace, your interactions must be recognized there. To show why, consider a distributed interactive simulation of the sort used in military training. If such a simulation is to work at all, there must be a synthetic universe into which all player attributes and actions are mapped. Supposedly, all players could factor in everyone's moves and initial attributes in their own unique way (for example, what you see as driving, others see as flying), but inevitably the result would resemble nothing so much as the argument of children: "you're dead," "no, I'm not," "yes, you are," "no, I'm not." So there has to be a master set of rules for any given space. Messages (that is, byte-strings) that do not accord to the rules are invariably rejected as meaningless, regardless of

---

[4] "Mutually accepted" is not meant to imply commonly understood. People may think the game has certain rules when, through simple misunderstanding or subterfuge, it turns out to have quite different ones.

[5] See Julian Dibbell, "The Unreal Estate Boom," *Wired*, January 2003, 11, 1, pp. 106–13.

how earnestly or maliciously put forward. The intrusion of a third party must be reflected in a change in the game's state; again, whether calculated centrally and broadcast or, instead, calculated individually in an identical way is secondary. Unrecognized actions or the actions of unrecognized parties do little harm except for perhaps clogging the lines. And, of course, not every player need be human; they can be machines.

Third, some aspects of cyberspace nevertheless tend to be persistent. A few rules of cyberspace, such as the laws of cryptography, derive from mathematics. Others are artifacts of well-accepted conventions (such as TCP/IP) or reflect the dominance of certain products in the marketplace (such as Microsoft Windows). One can construct a cyberspace without them, but most information systems adhere just the same. Such rules can come from many places, such as from those who write the software or from the community that maintains the environment in which the software runs.[6] So, while these rules remain constructs, they are constructs in which people have invested value.

Certain systems, as well, are persistent. There is, for instance, only one Internet, and it has certain conventions such as a hierarchy of routers as well as a set of recognized names corresponding to a set of recognized addresses.[7] But there are also internets (small "i") that use the same ubiquitous and richly supported communications protocols as the Internet but are not connected to the Internet. There are yet others, which are connected but in ways that make it very difficult for the innocent public or not-so-innocent hackers to get into them.

Even at a macro level, as Lawrence Lessig[8] has argued, cyberspace has nearly no inherent properties and only a few strong tendencies; everything else is imposed by those with the power to do so. The oft-cited aphorism that the Internet interprets censorship as network damage and routes around it has been used to imply that the inherent qualities of the Internet

---

[6] It would be harder to change unilaterally games such as Dungeons and Dragons, whose rules have evolved organically over time.

[7] To illustrate that even constructs have value that can be captured and traded, note the large amounts of money associated with certain domain names that serve as beacons in a fog of potential URLs. Nevertheless, the tendency to type "socks.com" in order to begin shopping for socks is being replaced by that of typing "socks" into Google.

[8] Lawrence Lessig, *Code, and Other Laws of Cyberspace*, New York (Basic Books), 1999.

have made free speech inevitable in that medium. But this runs up against the real-world constraints that governments such as China have largely imposed successfully on its Web users. There is, Lessig goes on to argue, a substantial capability to express social norms, hitherto reified only in legal code, as computer code to achieve roughly the same ends. To say that cyberspace is a "commons" or a "market" presupposes some expression in cyberspace of social norms and, in some cases, legal enforcement that permits commons or markets to function in real space. It does not arise from the nature of cyberspace or always come out in the same way. For instance, EBay, the online auction site, is a global market, and there are mechanisms (for example, a global reputation registry) that work on EBay to provide other or more efficient ways of enforcing commercial norms that exist in the physical world. But supposing such markets would work absent any mechanisms and social norms whatsoever is naive.

Fourth, cyberspace has separate layers, the conquest of each of which has vastly different meaning. Stated briefly, and discussed in much greater detail in Chapter 10, one can define three layers in cyberspace with their parallels in linguistics: the physical, the syntactic, and the semantic.[9]

The physical layer – including such things as wires, routers, and switches – is the foundation of cyberspace in the tangible world.[10] Conquest that takes place here could be understood in terms of physical control over the infrastructure – frustrated only by the ease with which most of the infrastructure can be replicated if necessary.

The syntactic layer reflects both the format of information in cyberspace and how the various information systems from which cyberspace is built are instructed and controlled.[11] As explained further

---

[9] Chapter 10 also discusses a fourth layer, pragmatics – essentially the intentions that lie behind the speech acts. Until such time – and it may be coming – that one can usefully impute intentions (or goals) to programs and machines, the pragmatic layer applies only to person-to-person interaction mediated through cyberspace. Thus "conquest" at this layer is very hard to define.

[10] It is not impossible to build a functioning cyberspace atop a biological stratum and use it to convey analog and/or fuzzy information (as today's nervous system does). All the software needs to know are the system's abstracted basic parameters (for example, how fast, how reliable, how ubiquitous, how much capacity).

[11] One of the great engineering successes of the TCP/IP protocol (and the Internet conventions that rest on them) has been to push the intelligence into the periphery of the system rather than concentrating it in the control infrastructure. J. H.

later in this chapter, the syntactic layer can itself be divided into sublayers; one canonical model, Open System Interconnection (OSI), identifies seven of them. Control here is often a matter of mastery: Can my knowledge of the rules overcome your knowledge to get machines to do what I, rather than you, want? And who writes the rules?

The semantic layer contains the information meaningful to humans or connected devices (for example, machine tools). Here the issue is one of influence: can I present to you a different version of reality that others take to be true?

So, conquest works differently at different layers. Physical access (that is, connectivity) does not mean syntactic access. Syntactic access does not mean meaningful semantic access. And semantic access does not necessarily result in meaningful change in what people believe about the world or even about cyberspace.

The layers of cyberspace may be likened to a party hall with private rooms. All these rooms are part of the same physical structure and they are mutually accessible, but that does not mean that what goes on in one room says much about what goes on in the next. To get into any one room may require a key (in cyberspace terms, knowledge of the network address and the password). Those who make their way in still have no guarantee of meaningful interchange with any of the participants. One may be simply ignored or not understood. Becoming a meaningful part of the conversation has three requirements: getting to the party hall, finding a key to the private room, and being accepted by those who are conversing. Some party rooms are better than others, in part because of better physical facilities (in cyberspace terms, faster connections, better data stores, more sophisticated support services, and so on). Some conclaves such as chat rooms are open to everyone.[12] Others, such as The Well (a Sausalito-based

---

Saltzer, D. P. Reed, and D. D. Clark, "End-to-End Arguments in System Design" *ACM Transactions in Computer Systems,* November 1984, 2, 4, pp. 277–88, also available at web.mit.edu/Saltzer/www/publications/endtoend/endtoend.pdf; see also David Isenberg, "The Rise of the Stupid Network" (www.isen.org). This is because the packets that carry the information payload also carry, embedded within them, processing instructions to the network. This instructions/content relationship has analogies to the syntactic/semantic relationship of human language.

[12] In practice, spaces such as America Online's (AOL) chat rooms are open only to AOL members and can exclude known abusers.

bulletin-board system [BBS] that predated the Worldwide Web site), are by invitation only. The more desirable neighborhoods in cyberspace are often so because they are better organized with more entertaining activities and intriguing conversationalists; others are more interesting because they permit certain types of business to be done.

Even, perhaps especially, Islamic (more technically, Salafist-Jihadist) terrorists hang out in their own neighborhoods to transact their "business." In some cases, notably when propagandizing for the masses, seeking recruits, or distributing Web materials, these neighborhoods tend to be public. In other cases, when mooting plots among themselves, Jihadist sites are more private; access is carefully revealed to known individuals. These are not rigid or even rigidly enforced distinctions. Sites have been penetrated by researchers who have figured out how to sound like a potential terrorist.[13]

It turns out that hostile conquest in cyberspace takes place largely at the physical and syntactic layers, while friendly conquest in cyberspace, because it has to do more with the exchange and encoding of knowledge, tends to take place at the syntactic and semantic layers.

1.2 Précis

The contention – that it is hard to control the world by hostile conquest in cyberspace but that the power available through friendly conquest merits attention – is developed in three parts followed by a conclusion.

Part I deals with hostile conquest in cyberspace. It starts with the premise that information systems exist to generate information, that information is used for decisions, but that humans are the agents of knowledge and decision making.[14] In other words, if one is to attack systems and so affect decisions, one must recognize the decisions are ultimately made by natural rather than artificial cognition. People, unlike

---

[13] Notably the Search for International Terrorist Entities' (SITE) Rita Katz; see Benjamin Wallace-Wells, "Private Jihad: The Woman Who Became a Freelance Spy," *New Yorker*, May 29, 2006, pp. 28–41.

[14] While acknowledging that systems are also used to make decisions and carry out actions automatically, the fundamental choice of where and how to put people into the decision loop is quintessentially a human one as well.

computers, have a great (if not always used) capability to learn, intuit, create, and adapt. They are capable of holding what the machines tell them at arm's length.

This helps frame, in Chapter 2, "Hostile Conquest as Information Warfare," the potential for and limitations of hostile conquest in cyberspace. Since the 1990s, when cyberspace came to the attention of DoD as a potential medium of conflict, actions in it have been considered part of a broader topic, information warfare. Since all military action was based on information, information warfare must therefore carry strategic weight. From afar, one could wage war and thereby take down entire countries – and all without showing up or even revealing oneself as the attacker. But this conclusion is based on conflating the information that people use to make decisions and the information used to control information systems. Attacks on information systems are not necessarily attacks on the truth of the information they hold or pass. A quick comparison to nuclear conflict suggests the absurdity of building up hostile conquest in cyberspace as strategic.

Further limitations on the usefulness of information, as Chapter 3, "Information Warfare as Noise," explains, come from making an analogy to information theory. If information is signal, then warfare on information can be envisioned as the creation of antisignal or noise. Error is believing something that is not, in fact, correct. Noise, though, is uncertainty that one knows that something. They differ. In the short run, bad information may be believed. In the longer run, successful information warfare may cause victims, whether man or machine, to regard all information with caution and draw upon multiple sources for confirmation. Noise complicates decision making but it is not necessarily a mechanism for control per se and hence conquest. Much depends on the noise-tolerance of the system being attacked; systems that deal with large quantities of questionable information (for example, financial markets that ride on rumor and reportage) fail in different ways and need different defenses than those that deal with small quantities of high-confidence information (as in command-and-control systems).

As Chapter 4, "Information Warfare against Defense Systems," relates, although it is impossible to say that information warfare would *not* work, so is proving with any confidence that it will work, much less that it will work even close to as well as a model might predict. There are many potential ways for hackers to get into a system and many possible ways to cause mischief once there. Nevertheless, there is no path to success that can be repeated at will, and success depends on largely unknowable, often minute, details of how target systems are constructed, manned, and maintained.

Indeed, even the premise that information warfare should be designed to destroy information is hardly obvious; it could be the reverse. Chapter 5, "Information Warfare against Command and Control," observes that people these days are more likely to find themselves not with too little but with too much information, albeit of a sort akin to low-grade ore that requires considerable smelting and refining to be useful. There is no such thing as a totally costless strategy for coping with information overload, and some coping strategies are downright neurotic. Perhaps information warfare is less a matter of destroying the other side's information, and more a matter of adding needlessly to it – even if this does not sound like conquest.

The logic of friendly conquest in cyberspace, the subject of Part II, depends on the willing, perhaps enthusiastic, assent of its victims. It is rarely something one can try on dedicated enemies, but it may be worthwhile with friends, neutrals, and those yet to become enemies. It may prepare or vacate the battlefield before the contestants make up their mind to choose sides.

Chapter 6, "Friendly Conquest in Cyberspace," illustrates the use of seductive appeal as it might work in cyberspace. To wit, one who controls a system may let others access it so that they may enjoy its content, services, and connections. With time, if such access is useful and assured, users may find themselves not only growing dependent on it, but deepening their dependence on it by adopting standards and protocols for their own systems and making investments in order to better use the content, services, or connections they enjoy. The harder it is for users to walk away from such a relationship, the more power a system's controller potentially has over them and the assets they have entrusted to that relationship.

If the bigger and richer the system, the greater the draw, then systems that span the globe should have the greatest potential for friendly conquest in cyberspace. Chapter 7, "Friendly Conquest Using Global Systems," walks through some of the issues associated with systems design and access for two potential global systems. One is a conglomeration of public and private remote sensing systems used to increase global transparency. The other deals with a universal registry of identification information (such as biometrics and biographies). In both cases, dominant systems can set the rules and make it harder to justify establishing systems based on competing ones. To the extent that such rules serve particular purposes, their influence is stronger because others have brought their systems into alignment with such rules.

It would nevertheless be misleading to argue that hostile conquest and friendly conquest are mutually exclusive. To the contrary, as Part III explains, friendly conquest makes hostile conquest that much easier.

Synergies between friendly and hostile influence may characterize the relationship between information systems and the individuals they cover. With every year, as more information is collected on each of us – and then indifferently guarded or even gleefully sold – the risk grows that those who do not have our best interests at heart can, through collection or theft, acquire large quantities of knowledge on us. This fact alone may help convert the techniques of persuasion from a wholesale to a retail enterprise. But, as Chapter 8, "Retail Conquest in Cyberspace," asks: what can they do with it?

Business, military, and other coalitions are coming to depend on the privileged exchange of large quantities of information. They can only do so if such exchanges are based on trust both in the information itself and on the protections that such information receives. Inevitably, as Chapter 9, "From Intimacy, Vulnerability," suggests, closeness exacerbates the vulnerability of partners to each other's designs and each other's capacity for fecklessness. Conversely, because the right kind of coalitions can be so valuable in global competition, opponents of the coalition may use information warfare to sow distrust within such coalitions so as to fracture or at least corrode them.

Chapter 10, "Talking Conquest in Cyberspace," draws explicit analogies between the layers of human language – phonology, syntax, semantics, and pragmatics – and their counterparts in cyberspace as applied to information exchange. This analogy is used to examine how (1) complexity helps blur the boundary between syntax and semantics and thereby facilitates hostile and friendly conquest in cyberspace, (2) the deepening elaboration of a semantic layer may facilitate friendly conquest in cyberspace, and (3) the potential development of a pragmatic layer may frustrate hostile conquest.

Many of these topics necessarily involve public policy, as Chapter 11, "Managing Conquest in Cyberspace," discusses. What is the relationship, if any, between the defense of any organization's share of cyberspace and the defense of a nation's cyberspace? How should a nation or world be protected from the deleterious effects of information warfare? Conversely, what strategy should the U.S. government, and not only DoD, follow in order to take the greatest advantage of cyberspace? Thanks to the influence of its software and the pervasive mechanisms of its corporate strategy, Microsoft is the pivot on which the contrasting versions of conquest are examined.

Appendix A, "Why Cyberspace is Likely to Gain Consequence," tells why cyberspace is becoming increasingly consequential – more ubiquitous, more complex, more globalized, and animated with more intelligence (of an artificial nature).

2

Hostile Conquest as Information Warfare

Information systems, the basis of cyberspace, are powerful and hence valuable. But they are also complex and hence often incomprehensible. As tools they should be subject to our mastery. But do we really control them? Even our stand-alone systems may arrive corrupted or may become corrupted through something they ingest. Connect them to the world, whether by yesterday's floppy disks or today's networks, and our fears multiply – and not without justification. Will these systems be available to do our bidding when asked – or will they sicken with viruses or drown in spam? Will information we entrust to our systems stay put or follow some Pied Piper out of town? Can we even be sure that the information they present to us has not been tampered with?

This combination of growing dependence and ever-shaky confidence in our control over information systems has given rise, since the early 1990s, to a new type of threat, and conversely and consequently a new branch of the military art: information warfare. Information warfare is often epitomized by hostile operations in cyberspace although, as explained shortly, it can take place in other ways.

This, the first of four chapters on hostile operations in cyberspace, builds an archetype of information warfare and then works backward to its epitome – computer network attack. Embedding computer network attack within a broader context of information warfare provides a sense of where it stands within the whole military milieu. Understanding hostile conquest in cyberspace to be information warfare also makes it clear exactly what its purpose is and what other ways exist to gain the same end. Finally, because information theory is already well grounded, one has a basis from which to build a theory of information warfare and hence conquest in cyberspace. Accordingly, the three chapters that follow this one focus on the following:

- A theory of information warfare based on information theory
- The use of information warfare against defense and other critical systems
- The use of information warfare to choke command and control

2.1 An Ideal-Type Definition of Information Warfare

Information warfare, as a term, has been used to encompass all forms of warfare carried out to affect an enemy's information. One monograph[1] described seven distinct activities that could be and have been called information warfare, each of which has enjoyed currency in some community. The first five describe practices in use, the last two hypothetical ones:

- Command-and-control warfare
  - against leadership
  - against communications
- Intelligence-based warfare
  - antisensor operations
- Electronic warfare
  - against sensors (for example, radar)
  - against communications
  - interception/exploitation
- Psychological operations
  - targeting leadership
  - targeting populations
  - targeting opposing forces
  - cultural struggle
- Hacker warfare (also known as computer network operations[2])

---

[1]　Martin Libicki, *What Is Information Warfare*, Washington (NDU Press), 1995.

[2]　Technically, computer network operations – the official term within DoD – is a bit of a misnomer. Information systems unconnected to networks can also be tampered with by insiders or can be corrupted by media that have been tampered with (for example,

- Economic information warfare
- Cyberwarfare
  - trial by simulated combat
  - cyberterrorism against individuals
  - semantic attack
  - war among cyber-beings à la William Gibson

Should such a capacious lexical tent shelter such a vast array of techniques with little in common but that they affect what information it is that people, notably but not exclusively enemies, know? What a daunting theoretical challenge it is to cover, in one treatment, computer hackers, cryptographers, electromagnetic wizards, media hounds, leaflet droppers, spies, drivers of airborne radar, special operators, bombers, and sharpshooters who target opposing generals?

The most egregious conflation, and one that still hobbles thinking within the Pentagon, lumps computer hackers and psychological operators together. If nothing else, they both stare at screens for a living – the former at computer screens examining computer script, and the latter at television screens examining news script. To add to the confusion, circa 1996, what was "information warfare" became "information operations" within the military, ostensibly to subsume activities, such as propaganda, that could take place during peacetime. Nevertheless, the term, "warfare" had much going for it. It had Anglo-Saxon clarity and served to remind everyone that two could play. "Warfare" connotes the Clausewitzian contest between two wrestlers each seeking advantage over the other. "Operations" somehow connotes doctors hovering over an anaesthetized body on a table. Alas, a military that insists on calling something an operation rather than warfare risks being surprised when the object of its interest rises from the table and hits back.

Little wonder, then, that well over a decade after the topic of information warfare broke out into the open[3] its conceptual underpinnings

---

floppies, as were common ten years ago). In extreme cases, systems may have been born corrupted. Yet, almost all of the mischief wreaked upon information systems is expected to come via networks of some sort, so the imprecision in the name is of limited consequence.

[3] During the 1980s and early 1990s, the very vocabulary of information warfare was highly classified, even though many of what were considered its components were discussed

remain weak and largely unsatisfactory, with fierce battles raging over neologisms and definitions. Despite the seminal pieces on the topic,[4] an early attempt at grand conceptualization,[5] and efforts to define the nature of strategic information warfare,[6] the very nature of the subject as a whole still defies theoretical treatment. It hardly helps when seemingly every new piece of hardware and software creates new capabilities; each new capability creates new vulnerabilities and new openings for mischief – hence new potential for information warfare and hence a new definition in practice. To give a small but telling example,[7] in the early 1990s, victims of most computer viruses (for example, ".exe" viruses) acquired them by booting them up from an infected floppy disk. Very few computers boot up this way any more. The next wave included "macro" viruses (for example, the Melissa Virus) in which an innocent-looking document contained an embedded series of commands. By 2000, the worm wave (including, for example, the Code Red worm) exploited the fact that many client systems (such as user workstations) actually contain server capabilities. All this took less than four years. This last wave is hardly likely to be the final one; viruses specific to personal digital assistants (PDAs) and digital telephones have started to

in the open literature. One reason for the high classification was the focus on strategic deception and perceptions management – efforts that would lose all force had they been revealed as such. By contrast, Hollywood has never had a problem dealing with information warfare. As far back as the 1983 movie *War Games*, proof of a character's high intelligence lay in the ability to penetrate computer systems. The plots of the three big summer blockbusters of 1996 – *Mission Impossible*, *Eraser*, and, especially, *Independence Day* – all hinged on some act of computer hacking. See also Thierry Breton, *Softwar* (English translation), New York (Holt, Rinehart, and Winston), 1986.

[4] John Arquilla and David Ronfeldt, "Cyberwar Is Coming!" *Comparative Strategy*, 12, 1993, pp. 141–65. See also Richard Szafranski, "Neocortical Warfare? The Acme of Skill," *Military Review*, November 1994, pp. 41–55.

[5] The first attempts to develop a grand theory for information warfare were generated by the late Tom Rona in the late 1970s (see Bruce Berkowitz's discussion of Rona in *The New Face of War: How War Will Be Fought in the 21st Century*, New York [Free Press], 2003). By and large, they were fairly high-level excursions into conceptual space. They did not dwell on computer network attack largely because the computers that existed were largely closed off from the outside world.

[6] See Roger Molander, Andrew Riddile, and Peter Wilson, *Strategic Information Warfare: A New Face of War*, Santa Monica (RAND), 1996, as well as Roger Molander and Peter Wilson, *Strategic Information War Rising*, Santa Monica (RAND), 1998.

[7] Robert Lemos, "Fast-Spreading Code Is Weapon of Choice for Net Vandals," http://news.com.com/2009-1001-254061.html, March 15, 2001.

appear.[8] Although it is hard to generate a cogent theory without due regard for what is and is not possible in a technical sense,[9] it is easy to be distracted by year-to-year changes in what is, in fact, possible.

For this reason, it may be more useful to set a foundation in first principles and then work toward a definition of information warfare, against which all forms can be compared – a Platonic[10] exercise with the definition being an ideal type and the instances being approximations. Indeed, the official[11] abandonment of the term "information warfare" has its silver lining. The term can be recycled and thereby assume a greater clarity.

---

[8] "The first serious outbreak of a mobile-phone virus in a company has been detected, according to security specialist F-secure." Tom Espiner, "F-Secure: Commwarrior Claims First Big Victim," http://news.com.com/2102-7349_3-5845021.html, August 31, 2005. Apparently, even opening up a malicious image sent via an e-mail on a Blackberry can disable a user's ability to view attachments. Joris Evers, "BlackBerry Users Face Security Threat," http://news.com.com/2102-1002_3-6016847.html, January 3, 2006.

[9] In the late 1940s, atomic scientists would argue (perhaps self-servingly) that those who did not understand the physics of nuclear reactions were therefore ill-equipped to discuss nuclear strategy – despite the fact that it was blindingly obvious what nuclear weapons did if not necessarily how. Those who still credit this argument would find it even more applicable in information warfare, whose effects are highly variegated and contingent on what one finds when entering someone else's information system.

[10] Ideal-type definitions can be contrasted with binary definitions. A good binary definition says what is inside and what is outside a category. A koala bear may resemble a polar bear but they are absolutely differentiated by their genetics, which, in turn, are a function of their ancestry. An ideal-type definition is one in which no real case need necessarily be 100 percent indicative, but there are many cases that realize the ideal to a greater or lesser extent. There is no sharp distinction between a hill and a mountain, although there are landforms that most people would classify as hills and not mountains, and vice versa. Binary definitions correspond to normal sets; ideal-type definitions correspond to fuzzy sets. Theorists of fuzzy sets argue, though, that normal sets are just a special case of fuzzy sets (see, for instance, Abraham Kandel, *Fuzzy Mathematical Techniques with Applications*, Reading [Addison-Wesley], 1986).

[11] Although the term "information warfare" has left the building (that is, the Pentagon), it can still be found in the writings of the National Security Council. The term appears in its white paper "The National Strategy to Secure Cyberspace" (President's Critical Infrastructure Protection Board, draft white paper, September 18, 2002) on pages 4, 9, 44, et al. The Air Force also uses "information warfare" to subsume both "information operations" and "information in warfare," the latter being all techniques by which the use of information is used to enhance warfare in the sense of physical attack and defense. Information warfare has also been used to describe military operations that are substantially enabled by information systems. See, for example, Stephen Blank, "Can Information Warfare Be Deterred?" *Defense Analysis*, 2001, 17, 2, pp. 121–38.

So, here goes: Information warfare is *the use of information to attack information.*

This definition immediately raises the question, what does it mean to attack information? We start with the premise that the purpose of information is to guide decision making, thereby classifying, by exclusion, all information that does not bear on decision making as entertainment. Unless information warriors are in it for the entertainment value – and most amateurs, indeed, are – they are perforce attacking information processes in order to mislead or misdirect decision making to their own advantage.[12] Insofar as the purpose of information is to make better decisions, the purpose of information warfare must therefore be to confound the making of these decisions, including those made by machines[13] – for example, by restricting access to files or by misdirecting flows in a oil refinery.

By extension, the purpose of, say, conducting information warfare on a command-and-control apparatus is precisely to reduce the victim's ability to command and control. The purpose of going after air defense radar is to reduce the ability of the radar to acquire, track, target, and engage aircraft. Attackers need not target any specific decision, although they might if they knew enough about what had to be decided. Attacking information therefore ought to influence the other side's decision-making processing to one's advantage: wrong decisions, late decisions, and decisions that, while perhaps good for the enemy, are also good for oneself (for example, the other side chooses to stop fighting).

Information, itself, can be destroyed or degraded in many ways. It can be erased. It can be misplaced or made hard to find. Information can

---

[12] The outcome of the affected decisions need not enhance the attacker's welfare directly. An attack may be made to frustrate the foe or to establish a basis for suasion – for example, "do this and I will not hurt you by attacking your information."

[13] The U.S. Joint Chiefs of Staff, *Joint Doctrine for Information Operations (JP 3–13)* of February 13, 2006, p. GL-9 (which can be found at http://www.dtic.mil/doctrine/jel/new_pubs/jp3_13.pdf) defines information operations as follows:

The integrated employment of the core capabilities of electronic warfare, computer network operations, psychological operations, military deception, and operations security, in concert with specified supporting and related capabilities, to influence, disrupt, corrupt or usurp adversarial human and *automated* decision making while protecting our own [emphasis added].

be *indirectly* degraded by being removed from its context (for example, a business card is left on the table as a reminder to call the number, but if it is reinserted into a stack of cards, the implied message is lost). The addition of contradictory or confusing information to the existing stock of information can also indirectly destroy the latter. Adding another message that looks as real but says the enemy is west can vitiate the value of a message that says the enemy is east. As a corollary, one can destroy or degrade information not by altering the information itself but by altering the credibility with which it is received. Conversely, and one hopes not confusingly, one can degrade information by inducing people to give certain assertions *more* credence than they deserve.

Prior to efficient duplication, most information that did not come from a printing press existed only in the original. Thus the *direct* destruction of information invariably was possible through the destruction of the medium on which it was written (for example, by burning papers). Techniques of physical destruction still come in handy. Destroying a central computer tends to destroy the information held by that computer – if such information is not backed up elsewhere. That caveat suggests why efforts to destroy someone else's information these days make little sense if directed against someone making even modest efforts to ward off such a threat. Storage is extremely cheap and it continues to get cheaper. The dollar that could buy a megabyte of storage in 1993 could buy a gigabyte's worth ten years later. Companies routinely back up the information held by users in specific servers and many take pains to create offsite repositories to ensure that they can carry on in the case of disasters, natural and otherwise. Storage area networks exist to ensure that users are blissfully unaware, even momentarily, of any disruption whatsoever. Clearly, such pains are harder to take in an austere battlefield environment, but the same basic trends appear there also. There is also the everyday tendency for files to multiply in the course of being circulated and modified. A great many of them could probably be painstakingly recovered even in the absence of contingency planning. Indeed, information assurance is plagued by the opposite problem: there are so many copies around that it is difficult to ascertain which is most current or whether or not one of them may have shuffled out the door.

Conversely, the alteration of paper copies without detection requires a fair level of skill, while alteration of computer files – for which the term "original copy" has little meaning – is not difficult, as such.[14]

A greater parallel can be drawn between historic and modern ways of destroying information conduits. Yesterday's messengers and telegraph lines are today's fiber-optic lines. Destroying the fiber-optic lines that connect the computer to the field can introduce a growing divergence between what the field should know and what the field does know, and vice versa. It is like information warfare in the sense that it degrades the correspondence between the state of information at the field and headquarters.

Information in cyberspace can be further differentiated from information as traditionally stored in the sense that modern networks can be made accessible from indefinite distances in ways that paper copies cannot be. Guard the perimeter, vet the employees, and buy adequate firefighting gear and you have solved most of the problems of keeping information on paper intact, if not necessarily organized. But for more and more networks, there *is* no geographical perimeter as such. Connect something to the Internet and, as real-time pictures from the red planet revealed, it is accessible literally from as far away as Mars. That means that it is also accessible from caves in the mountains of Pakistan. True, physical proximity has its advantages; it does help in social engineering (the art of conniving people into giving up their passwords or revealing other systems secrets), dumpster diving (extracting valuable information from carelessly discarded garbage), or intercepting signals from wireless networks. But putting physical distance between you and your enemies is no panacea in cyberspace.

In contrast to past media, today's information systems themselves use information, to manage information. The information that such systems use includes such things as programs and other files; the information that they manage is the content that people use as a basis of making decisions (plus programs and files for systems management).

---

[14] Electronic documents can, however, be made quite hackerproof. Systems could time-stamp and digitally sign documents. If the server's logic is immune from tampering (for example, if it is hard-wired) and the private signing key is never revealed, a newer copy would be impossible to manufacture without its illicit provenance being detected.

That information systems use information to operate creates a major source of confusion: the distinction between *attacks on information* and *attacks on information systems*.[15] The first is content – the stuff upon which decisions are made. The second is management – the stuff that facilitates making decisions at all. Both can be rendered in ones and zeroes, but they operate at different levels.

Because information systems, themselves, rely on information to know what to do, one way to disrupt information systems is to corrupt or at least hide the information that they rely on to function correctly. Corruption can also cause computer information to wind up in the wrong hands (for example, by convincing the computer that a particular mailing address represents a privileged user rather than an imposter).

At one level, computers see information that is used to make decisions being very similar to information used to create, manage, and manipulate such information; it's all bits. Both can be attacked in similar ways. Nevertheless, because software may be used to guard different information in different ways (for example, encrypting some but not all of it), the two types of information may not be vulnerable in identical ways.

Some important features differentiate the *effect* of attacks on information going to people from that of attacks on information going to systems. People use information systems to get information. If the information systems are disrupted or disabled, their access to information and information services may be denied, but the integrity of the information they have may not necessarily be affected. For reasons discussed in the next chapter, it is usually more difficult to corrupt information going to humans (such as semantic content) via computer network attack than it is to corrupt information that help run information systems (such as syntactic content) – at least at this stage in the evolution of computers. But confounding the systems that manage information is a far cry from confounding decisions based on the information they manage.

---

[15] Note the easy confluence of the two, from JP 3–13: "Information operations involve **actions taken to affect adversary information and information systems** while defending one's own information and information systems" [emphasis in original text] (p. vii). Later it continues: "'Computer Network Attack' is defined as operations to disrupt, deny, degrade, or destroy information resident in computers and computer networks or the computers and networks themselves" (p. I-9).

So, those who would attack the mind of the enemy by attacking information systems may succeed in making much of what the enemy knows inaccessible, but not necessarily by changing the content of the information itself. It is akin to claiming that one has successfully messed with someone's mind by giving him or her a headache. Taking down someone's information system can make it hard for people to make decisions. It can make it impossible for devices, such as ATM machines, to make decisions, such as to whom to dispense cash. But one can also make human decision making difficult by inducing physical stress (for example, from noise, discomfort, or darkness).

2.1.1 Control at One Layer Is Not Control at Another

Analyzing information systems in terms of their physical, syntactic, and semantic layers as laid out in the first chapter may help indicate what their conquest means.[16]

The physical layer consists of the various means that permit the circulation of bits – whether reified as radio-frequency (RF) energy, electrical signals, or photons. Here one sees wires, routers, computer hardware, ground terminals, antennae, satellites – the works. Networks can be and are attacked at the physical layer, largely by physical means, whether through hard kill (breaking the boxes) or soft kill (frying the electronics). It is possible to damage the hardware of a network through the injection of information, but there are only a few opportunities for doing so (for example, inducing the odd software bug that causes hardware to cycle in deepening ruts or creating false fault conditions that cause people to mismanage hardware).

The syntactic layer consists of the various instructions and services that tell information systems what to do with information. Computer syntax may be said to encompass operating systems (OSs) and applications. Network syntax clearly includes routing, but also access controls and security, directories, utility servers, and commonly used databases.

---

[16] Chapter 10 discusses a fourth layer in terms of pragmatics, but pragmatics, as such, is a very uncommon feature of today's systems.

Generally, the syntactic structure of a system is attacked via computer hacking.

The semantic layer contains a system's information content.[17] This information may be stored (for example, in databases) or circulating (as in messages). The more essential, and often-missed, distinction is between information contained by a system and the set of beliefs, perceptions, and plans held by people who use the system. Ultimately, command and control is – and for a long time will be – about people: what they think, what they say, what they commit to, and how they act. There are many and various ways to affect the thinking of a person, such as through the manipulation of evidence, the arguments of intermediaries, or changes in information upon which a decision is based. It is an exaggeration to argue that the last is the most effective of the three – even if the dictum "you cannot argue with a book" is evolving into "you cannot argue with a computer." A great deal depends on whose beliefs one is trying to influence and what these beliefs cover. One may argue that while the last avenue is the least efficacious of paths, it is the only one actually open to information warriors – *and therefore the only one open at all*. But beware of seeing nails, because one's tools are limited to hammers.

Because different layers have different functions, what is influential at one level may be nearly irrelevant at another. The commander whose computer sits on a phone line at the end of the network has considerably more influence on what people know to do (the semantic level) than the

---

[17] The syntax-semantics distinction of systems is an extension of the distinction present in human language. Semantics is the meaning of words, and syntax is their arrangement (and, in English, also the arrangement of word roots, prefixes, and suffixes). In Internet Protocol (IP) networking, packets have headers/footers and content. The former indicate how a network is to process information; content is what the network is passing. IP networking (by contrast to voice telephony) uses in-band signaling, just as human language does. This helps explain why the syntactic integrity of information flows can be defined in terms of whether packets are sent and processed correctly; much of the information to do this is embedded in its packaging. These, in turn, determine the proper application of network functions such as addressing, authentication, routing, encoding, error correction, and encryption – the everyday workings of computer networks. The semantic content of information may be judged by whether it is correct (for example, whether is it self-consistent and comports to the real world). It is at the semantic level where meaning lies. Thus it is the level at which deception operates against people and other logic processing units. At the syntactic level, deception operates against networks and data-processing units.

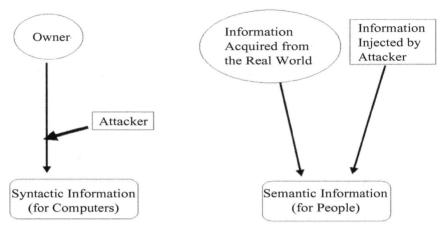

Figure 1. Attacks on Systems Information Compared to Attacks on Information for People

systems administrator who runs the routers that hub the fiber-optic lines at the network's core. But the two *could* sometimes sit in the same place. The switch that utilized stored data to route signals may also physically sit at a junction of many links. The database that stores who has what privileges may also provide information about who enjoys what trust levels within the organization. Ultimately, however, it is the flow of information into decision making that determines what effect subversion has. An automated process is much more vulnerable to attacks at the physical and syntactic layer than is a man-in-the-loop process. Conversely, a combination of stress and poorly designed man–machine interfaces may lead to human mistakes that automatic systems may avoid.

Figure 1 illustrates a key difference between an attack on syntactic information (that is, information used by computers) and an attack on semantic information (that is, information used by people). In the former, depicted on the left, the owner[18] would exert control over its computer systems through embedded instructions (such as programs and files) as well as real-time intervention. The attacker seeks to insert its own instructions, blocking the owners' instructions as necessary. In the latter,

---

[18] Here, the owner is defined as the person that controls the computer. In legal terms, if the computer is leased, the actual owner may be someone else entirely – an irrelevancy for this discussion.

depicted on the right, the owner is the decision maker. Owners rely on what they know (which is very hard to change) plus new information from the outside world (which is very hard to block in toto). Attackers would insert false or low-value information to affect this state of knowledge.

The important qualification imposed on the definition of information warfare – that it should comprise attacks on information *with information* – is entirely consistent with computer network attack and, as such, differentiates attacks on historic media (such as burning files) from attacks on current media (for example, persuading a computer to erase or alter its own files). Indeed, although the information held by computer systems can be destroyed (but there may be other copies around), their violent alteration is nearly impossible.[19] When applied to information systems, the randomization of matter that characterizes physical destruction is almost always recognized as the substitution of gibberish for information. Nevertheless, except for very stupid computers, such gibberish does not become the basis of subsequently misinformed decisions – although it may keep computers that need true content from working.

2.1.2 Applying the Ideal-Type Definition

How close do the various claimants to the moniker of "information warfare" come to this ideal-type definition?

Hostile operations in cyberspace, or computer network attack, and its converse, computer network defense, fit the definition very well – as long as one carefully distinguishes between attacks on information that feed decision making and attacks on information that feed information systems. Computer network attack relies on the insertion of information into or at[20] an information system to do harm. This has important ramifications, not least of which is that one cannot force another system to take the information. Yet, since information systems are built to accept at

---

[19] Physical destruction of computer data, however, can cause victims to doubt the integrity of the information they do hold if they are not sure if disconfirming information has been deliberately destroyed to strengthen initial conclusions artificially.

[20] "At" refers to attacks that disconnect systems from the rest of the world.

least some information, deception alone may suffice to permit a computer network attack. The systems' designers generally intend any such information accepted by a system to be useful to its correct functioning, or at least benign; otherwise the system would reject the information. Computer network attack plays on the presumption that the system "believes" that it is getting information it is designed to get and not information that harms its owners. Needless to add, not every system is designed as such; many are connected with insufficient thought about what all the interconnection will do to each part.

Computer network exploitation, the art of extracting information from a system against the will of its owners, is close to but not identical to information warfare as ideal type. Systems are built with rules to govern which information is passed to whom. Such rules assume the integrity of their own flow control systems and the information passed to their control systems for evaluation. Electronic eavesdroppers often bungle the workings of such control systems by introducing harmful information into and thereby inducing error in the mechanism (for example, by causing it to confuse authorized and unauthorized users). To stretch the point, computer network exploitation destroys information but only in the sense that it introduces a gap between information about what the control system is supposed to do (keep information out of the wrong hands) and what it is really doing (sending it out to the unauthorized). Unlike computer network attack, where a systems failure signals that something is wrong, successful exploitation can be a quiet affair; barring discovery of a leak or the discovery of errant code (often by accident), it can go on undetected and hence uninterrupted for years.[21] So, computer network attack and exploitation are close cousins. Both use similar tricks to get into information systems, and therefore call on similar skills.

---

[21] A venerable uncle of computer network exploitation, codemaking and codebreaking, can also be associated with information warfare. Here, ironically, the codemaker is trying to make real information look like gibberish while the codebreaker is trying the reverse. Codebreakers are trying to filter – more accurately, transform – what looks like noise into signal. In steganography (such as putting meaningful information in the least significant bits of a picture's pixels), codemakers are trying to hide real information within what are ostensibly the details of other information.

Both subvert control systems. But their intent is different. So is their legal treatment; destruction of information is more likely than eavesdropping to be perceived as an act of war.

As for physical attacks on information systems, the ends are similar but the means differ. Physical attacks can include destruction. They can also include the sorts of disruptions that could be produced by high-powered microwaves, electro-magnetic pulses (EMP), and other forms of violence against electronics. As noted, redundancy is getting cheaper; thus physical attacks are more likely to disrupt information processes than destroy information itself.

Electronic warfare could also be called a form of information warfare and it is, in many ways, close to the ideal-type definition. In one form of electronic warfare, jamming, attackers send out signals of their own to make it difficult for the defender to read intended signals (which could be either communications or radar reflections). In effect, the information content that is flowing from source to recipient is being degraded or completely destroyed even if the energy content is unaffected. Other aspects of electronic warfare create false signals to fool the victim's receivers. In that sense, false information "destroys" true information. Yet electronic warfare is not exactly information warfare. With jamming, the weapon of attack is RF energy rather than information per se. Energy becomes information only after it has been meaningfully processed.[22]

Classifying deception as information warfare rests on the sense that information is destroyed to the extent that people who have a true perception of the world are led to a false one (as is also true for certain types of electronic warfare noted previously). Much deception works by reinforcing false knowledge. Thus, the classic case of the Allies persuading Hitler to think that they would try to breach the Atlantic Wall at Calais rather than in Normandy (as they did) fits an ideal-type definition of information warfare. It was generated to reinforce Hitler's misinformation and inhibit him from reacting with full force to the D-Day invasion by leading him to believe that the Normany landings were a feint. Deception can

---

[22] With jamming, it matters little whether the attacker sends out noise or something that the victim could have but did not interpret as a signal.

range from operations focused to produce a certain effect to those that merely seek to change one fact or create a blizzard of facts in ways that obscure the one true fact.[23]

It is nevertheless a stretch to identify all psychological operations with information warfare. Humans are imperfect[24] thinking machines; they let emotions and other psychological needs get in the way of logic when making decisions.[25] Thus a good share of what is required to get people to make bad decisions is to play on their psyche. Fear and panic affect decision making; so do misplaced affection, gratitude, and the confidence that the actions of others can be predicted by evaluating their sincerity. Once people make up their mind, especially if they had to fight others to establish or justify their point of view, they are unlikely to change it without compelling evidence, and perhaps not even then. Con men, advertisers, and media consultants all understand this fact. Nevertheless, if one can subtract these emotional and psychological factors, many features of psychological warfare do fit the ideal-type definition of information warfare.

Figure 2 is a bull's-eye chart that depicts how close the various pretenders to the throne of information warfare come. Computer network

---

[23] Sometimes, though, repeating a fact in various permutations can create the kind of noise that, in effect, destroys information. Assume that a state seeking nuclear weapons built a nuclear facility. If identified, it would warrant destruction. From space, the facility is not particularly unique; there are many others like it. So merely finding it would prove very little. Among the many people who know what it is, someone decides to warn the world. Assume the whistleblower is anonymous to both sides (something quite possible in cyberspace). At first glance, attribution is not necessary for the information to matter. Even a no-name source can garner attention by naming a place and providing telltale details to look for; thus satellites can focus on the site and reveal the truth. The proliferator, however, suspects that the whistle is about to go off but does not know by whom. To protect itself, it engineers thousands of authentic-looking leaks. The true leak is lost in the noise; it no longer helps focus the search – unless it can be filtered out and forwarded. To convince others to take the leak seriously, the whistleblower could authenticate himself and his bona fides to the other side (by revealing himself as a famous scientist, for example) or provide the kind of detail that is hard to invent or cannot be divulged without giving the truth away (for example, the divulged detail might provide a nice explanation of something hitherto mysterious).

[24] Many experimenters have demonstrated that people make irrational choices (by economic standards), especially when it comes to judgment under conditions of uncertainty. See, notably, the articles in Daniel Kahneman and Amos Tversky, *Choice, Values, and Frames*, Cambridge (Cambridge University Press), 2000.

[25] See Robert Cialdini, *The Psychology of Influence*, New York (William Morrow), 1984.

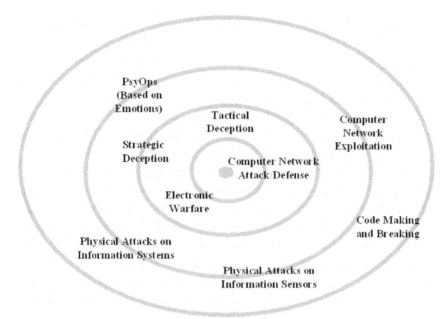

Figure 2. How Close Do Various Forms of Information Operations Come to a Canonical Definition of Information Warfare?

attack lies closest to the center. Huddled nearby are electronic warfare and psychological operations carried out by deception. Farther out lie computer network exploitation and the physical destruction of information systems. The figure shows how hostile conquest in cyberspace can exemplify information warfare, although examples and analogies drawn from the world of electronic warfare or deception are also apposite.

2.2 There Is No Forced Entry in Cyberspace

Attacking, exploiting, and defending information systems *successfully* depends much more on being able to understand their internal construction, characteristic features, and likely fault modes than it does on the type of physical features that play a larger role in other media of warfare such as the oceans. A knowledge of operating systems, failure modes, resource contention mechanisms,[26] cryptography, and other highly technical matters spells the difference between warranted and wasted effort. Because such attacks will have consequences in the real world, it is also important to understand the relationship between what happens when information systems are altered and the decisions that are therefore changed as a direct or indirect result.

A closer look at how information systems work indicates some of the basic mechanisms through which effects take place. It also suggests why there are limits to *what kind* of effects that they can have, even though there is no inherent limit to the *degree* of mischief they can cause. Information systems, after all, are controlled by the information they get. It should, one would imagine, be possible to control the information systems of others by giving them certain types of information. So, it would seem that warfare in cyberspace, for instance, should be about the control of a medium no less than warfare in any other physical medium.

To begin with, both computers and networks are built upon operating systems. They govern access to resources (such as files, screens, networks, and each other) and therefore contain the rules upon which access is based. These days, computer operating systems (such as Linux and Windows NT) tend to be network operating systems as well. This was not always true of past systems such as DOS, which was exclusively for stand-alone computers, or Novell NetWare, which was designed solely for networks.

In a network, an administrator assigns privileges to users and processes (such as applications) for reasons of order and security. Privileges range from the ability to access devices (such as a printer) to permission to alter files on servers or even other clients. Attackers often work their way to control by seeking increasing levels of privilege. They pose as users first. Then, if successfully acknowledged as users, they seek to subvert control processes to gain root access and the ability to read and rewrite every file. Root access permits attackers to read, alter, or destroy not only system files but the many client-side files as well. In effect, root access *does* leave its possessor with a certain amount of control. Those less adept

---

[26] Two processes contending for processor time can result in what is called a race condition, which can give rise to a security vulnerability.

at getting into specific places or less worried about collateral damage can flood target systems or release a menagerie of viruses through hapless hosts. The February 2000 E-commerce attack was of the first type; the May 2000 I-Love-You virus[27] was of the second.

What helps slip attackers past guards is that the control systems of cyberspace are complex, arcane, generally opaque, and brittle. They do not react well to the unexpected – which is to say, conditions for which they are not specifically programmed. Small coding errors in software can lead to major vulnerabilities if defenders fail to notice them.

The most fundamental defense is therefore solid software, but, in practice, users need to install defenses against the possibility – nay near-certainty – that software is anything but solid. Three (among many) types of defenses may be noted. The first is access control; most systems use passwords but cryptographic access controls are far more resistant to attack. These not only keep outsiders from assuming the privileges of insiders but, when used to support digital signatures, inhibit insiders from blithely messing with system controls. The second type of defense is the ability to separate benign from malign messages between components in a system; this also helps keep insiders from assuming the potential harmful privileges of systems administrators. Finally, there is the ability to exploit physical ownership to secure virtual systems; they range from air gapping (total electronic separation of networks from the outside) to the use of unalterable storage media and manual overrides such as the ability to shut down a network by "hand."

Digital signatures,[28] in particular, may well assume a growing role in protecting systems. They are algorithms applied to messages that

---

[27] This virus was, in effect, an automated chain letter. The recipient received a copy of an e-mail from a friend. The e-mail had an attachment that promised to be a note of affection but was really a Visual Basic program (technically, a script). Unbeknownst to the recipient, the program looked up the e-mail addresses of everyone who ever sent the recipient any e-mail; it then mailed itself off to those addresses (which is why those recipients thought that they are getting mail from someone they knew). The cascading series of e-mails affecting tens of millions of people one May day in 2000 clogged and thereby disabled thousands of workplace systems.

[28] For a general reference see Dorothy Denning, *Cryptography and Data Security*, Reading (Addison-Wesley), 1982; Bruce Schneier, *Applied Cryptography: Protocols, Algorithms, and Source Code in C* (John Wiley and Sons), 1995, especially pp. 483–502.

are impossible to generate except by someone with a particular private key. Any illicit change in the underlying message without a correspondingly nearly-impossible-to-calculate change in the signature string can be mathematically proven to be an incorrect message. For such a system to work, (1) the message, the algorithm, and the private key must all be uncorrupted when going into the digital signature algorithm; (2) the private key must remain private to the user; and (3) the private key must be reliably associated with the user. When properly used, digital signatures help ensure that all messages on the system are authentic, but only if these signatures are used carefully – and thus only if the targets know in advance that they need to use them to address the threat environment. One problem with using general-purpose computers for digital signatures is that such computers can be reprogrammed on the sly. If the programming is corrupted, so too might be everything it sends. A more secure device would feature hard-wired programming, accepting input by hand and outputting a provably authentic message. As long as certain tenets are followed, the system should be considered sufficiently secure.[29] In this way, an austere command-and-control system can be made practically impervious to the imposition of bogus messages. The limitations of such a system are easy to relate. The presumption that every message, in alphanumeric form, must pass before human eyes (and better yet, through human fingers) for validation before being signed limits its use as a general office automation tool. Associating a person with a device becomes complicated when the number of devices grows large. For this reason, while such a system may be useful for a limited population of

---

[29] To protect the user if the device falls into the wrong hands, it helps if the device requires a correct PIN to work, self-destructs if too many PIN numbers are entered, and uses a PIN that has enough characters that only the user knows, and which cannot be extracted from or altered in the device itself. If so, stealing the device is unlikely to permit the thief to generate a false message that passes scrutiny. Algorithms exist that decompose the signed message based on the recipient's public key and determine whether the message is authentic. That leaves the problem of ensuring that the list of public keys is valid; this can be solved by having devices accept only those changes to the public key list that are themselves signed by a single (or single and backup) source. If the number of new authorized message generators in an overall system is few enough, changes in the authorization list can be delivered on hard copy and then retransmitted under the authority of a central source. Direct (perhaps personal) authentication should work to ensure that people (and devices) are correctly matched up with a specific public–private key pair.

insiders, it may not scale well and thus would not be not as reliable if used to collect information from outsiders.

Because the elements of a computer network are under the ultimate control of the owner, one can confidently conclude that *there is no such thing as forced*[30] *entry in cyberspace*. If a destructive message gets into a system, it must be entirely across pathways that permit such a message to get through. Some pathways are deliberate design choices and have been inadequately guarded. Some arise from defects in the software. Some may be put there by attackers themselves after they have found the first pathway – but that pathway must preexist one way or another. Since computer systems developers rarely seek mischief as a design feature, most computer network penetrations involve some sort of deception[31] and can therefore be frustrated to the extent that deception can be unmasked.

Unfortunately, the tenet that everything is ultimately under the user's control must take due account of trends *toward* commercial operating

---

[30] "Forced" should be understood as it would be in the physical context (that is, a test of strength applied by each side to the same point). The legal context – that which a person is compelled to do under the law – is something else entirely. To spur competition, deregulated industries in this country (for example, telephony) are mandated to provide colocation and fairly priced interconnection to competitors. In theory, this can create vulnerabilities. Whether incumbent system owners can refuse or regulate such interconnection on cybersecurity grounds is ill-tested. AT&T damaged the presumption that systems owners could make such determination unilaterally when the Federal Communication Commission's *Carterphone* decision rejected the corporation's claim that customer-provided modems would harm the integrity of the Bell System.

[31] An important, but instructive, exception to the rule that computer network attack inevitably involves deception (or entry of some sort) is the flooding attack: sending a system so many messages (or packets or connection attempts) that others cannot get in because the entryways have been deliberately overcrowded. Simple flooding attacks are not new (for example, telephone switchboards can and have been deliberately flooded), but they are usually obvious, cannot last forever, and tend to point back to the original source fairly readily. A simple flooding attack requires the attacker (or attackers, collectively) to have more bandwidth than the victim does; this inhibits such attacks from being a tool of the weak against the strong. The February 2000 distributed denial-of-service attack on E-commerce sites, a more complex flooding attack, used viruses to draft unwitting third-party machines (henceforth termed "zombies") to participate; as such, it involved deception, at least against these third parties. To exacerbate the effects, an attacker may send a system incomplete packets, banking on the ultimate victim's system to keep each one in memory waiting for more information; if enough packets are sent, the router's memory fills up, knocking it out of service. This also involves deception in that the convention that mandates holding packets in memory assumes that bad packets occur randomly (and rarely) rather than deliberately.

systems[32] and *in* commercial operating systems – developments that are not under most users' control. These trends are here to stay. After all, using commercial operating systems and commercial software in general, as opposed to building one's own, makes sense from several perspectives. Everything one would want to see in a modern operating system is already there – and then some; indeed, the "and then some," unfortunately, is the source of considerable grief. It has been tested and (mostly) debugged. It is very inexpensive; single-processor licenses range from free to a few hundred dollars. Millions of people know how to work with it. Those who build their own OSs can rarely find third-party applications, hardware drivers, or easy ways to exchange documents with other applications (Web files being a notable exception). It is also getting harder and harder to work with the rest of the universe without commercial software. Even if institutions can afford to run only a small set of known applications that use a delimited set of function calls and have little need to exchange documents (other than Web files), they do so at the risk of having little room to accommodate unexpected demands or opportunities. They lock themselves out of using most new technologies, which invariably come only in forms compatible with commercial operating systems. So this is decreasingly feasible even for DoD – although open-source (and thus highly reconfigurable) Linux may provide a possible path to very secure systems.[33]

Users thus have little effective choice but to ride the trends *in* commercial software. One is greater complexity, which creates more places for security-undermining error. Others include remote maintenance (such as a software company updating the software on a client machine), always-on operations, and radio-frequency network connections (as reflected in Wi-Fi, Wi-Max, Bluetooth, and 3G Internet). These latter connections, in particular, reduce the amount of security otherwise present in systems

---

[32] This includes sufficiently well-distributed open-source systems such as Linux and the BSD version of UNIX.

[33] The National Security Agency signed a $1 million plus contract to develop an ultrasecure version of Linux. NAI Labs also received a $1.2 million dollar contract in 2001 to develop security extensions to the open-source FreeBSD operation system. Stephan Shankland, "NSA Funds Work to Thicken Linux Armor," http://news.com.com/2100-1001_3-255541.html, April 9, 2001.

under lock and key. When information is transmitted over the airwaves, even pulling out the wires will not secure a system.

Although making mischief is easy, permanently damaging a system through conquest in cyberspace is not. Novice computer users are coaxed into interacting with computers by being assured that they cannot damage hardware by hitting the wrong keys – which is true enough to be common wisdom. More frequently, information and/or programs are irrevocably destroyed only when insufficient effort is taken to back them up onto unalterable media (such as a write-once DVD-ROM); it is far easier to corrupt information as it is being processed or stored. Although systems can be returned to their original state, one often loses whatever adaptations and improvements have transpired since then. Information can also be illicitly copied throughout the process.

There is also a big difference between disabling a system temporarily and doing so for any great length of time. Evidence from widespread attacks (such as the Morris Worm, the AT&T/DSC phone outage, Melissa, the I-Love-You virus, the Code Red worm, and the Slammer worm) suggests that skilled systems administrators can often restore network performance within forty-eight hours – and inhibit further instances of the same techniques. For instance, every large server that was taken out of commission by the February 2000 distributed denial-of-service attack returned to service within three hours.[34]

Thus, any fair assessment of computer network attack must conclude that there are many available defenses, even if some are expensive and complex. Furthermore, certain types of damage are unlikely to result from even a successful attack except against poorly engineered systems and hapless owners. All this limits the kind of damage that even successful computer network attacks can have.

2.3 Information Warfare Only Looks Strategic

As societies everywhere become more dependent on information technology, can information warfare, despite everything so far noted, rise to

---

[34] Greg Sandoval and Troy Wolverton, "Leading Web Sites under Attack," http//www.news.com/2100-1017_3-236683.html, 0-1007-200-1545348, February 9, 2000.

the strategic importance enjoyed by nuclear warfare? Can it defeat the center of gravity of opposing societies?

Some have implied or stated as much. Lieutenant General Kenneth A. Minihan, while director of the National Security Agency (NSA), put protecting the nation from cyber-attacks and taking the offense with information warfare on the same level as protecting the nation from nuclear attacks: "Information warfare poses a strategic risk of military failure and catastrophic economic loss and is one of the toughest threats this nation faces at the end of this century."[35] James Adams observed in *Foreign Affairs,* "The private and public sectors together now form the front line of twenty-first century warfare and private citizens are the likely first target."[36] Even analysts at the normally sober Computer Emergency Response Team (CERT) offered, "[Computer attacks] would likely cross boundaries between government and private sectors and if sophisticated and coordinated, would have both immediate impact and delayed consequences.... Ultimately an unrestricted cyber attack would likely result in significant loss of life as well as economic and social degradation."[37] Finally, as analysts Ryan Henry and Edward Peartree, have observed in their survey of IW:

> The more radical of the theorists predict that information warfare will not only provide dominant awareness of the battlespace; it will also allow us to manipulate, exploit, or disable enemy information systems electronically. The intent here evidently is to knock an enemy senseless – literally – and leave him at the mercy not only of conventional kinetic attack, but of psychological operations aimed at controlling his perceptions and decision-making abilities. Public opinion is to be shaped, leaders will be cut off from citizens, and the mind of the enemy will be directly penetrated and his strategy defeated. In the ideal case, all this will occur bloodlessly, fulfilling Sun Tzu's goal of victory without battle. At least that's the theory.[38]

If so, can strategic information warfare be *usefully* likened to strategic warfare, which as far as major powers are concerned, is tantamount to

---

[35] James Bamford, *Body of Secrets,* New York (Random House), 2001, p. 422.

[36] James Adams, "Virtual Defense," *Foreign Affairs,* May/June 2001, pp. 98–112.

[37] Available at http://www.nato.int/docu/review/2001/0104-04.htm.

[38] Ryan Henry and C. Edward Peartree, "Military Theory and Information Warfare," *Parameters,* Autumn 1998, pp. 121–35.

nuclear warfare? Are the latter's metaphors, assumptions, and vocabulary a useful theoretical touchstone?

Almost certainly no – the two are really as different as fire and snow-flakes. Indeed it is a measure of how few other threats America faced, at least prior to September 11, 2001, that we could even entertain such a thought with any seriousness.

Nuclear war creates firestorms, destroying people and things for miles around. By contrast, even a successful widespread information attack has more the character of a snowstorm. It is not that snowstorms are smaller: in terms of lost income, the one that socked in the eastern seaboard in January 1996 approached $10 billion, a level no U.S. accidental firestorm[39] has even reached. But the effect of snowstorms, apart from a few random heart attack and accident victims, is entirely temporary and rapidly over. Furthermore, although U.S. cities differ only modestly these days in their susceptibility to firestorms, there are great differences in their suscepti-bility to snowstorms. The six inches worth that would send Washington, D.C., skidding to its knees would scarcely be noticed in Buffalo, New York. Equipment inventories, emergency management techniques, traf-fic regulations, snow removal zones, as well as personal expectations and coping strategies play a large role in determining how well cities with-stand flakes. Individual strategies, such as keeping the larders well stocked and (for those with the right jobs) hauling enough work home to remain productive can see people through. So it is with an information attack. Clearly, some systems are impervious to techniques against which other systems lie prostrate. Institutions vary greatly in their ability to overturn the results of such mischief and keep operating while their systems are being restored.

People impose nuclear weapons on others, but, as noted, there is no forced penetration in cyberspace. Hackers have little extant ability to create entry paths – only to exploit them. Information warfare, as noted, is strongly related to deception at one level or another (such as password stealing or Trojan horses), and the capacity to be deceived is clearly an

---

[39] The most destructive accidental firestorm in recent years was a fire in the Oakland hills, which destroyed over four hundred houses, killed twenty-five, and did roughly $1 billion worth of damage. Fire from the *deliberate* attack on the World Trade Center was far more deadly and costly (even when the damage from fire alone is considered).

inherent quality of a system (or humans, for that matter). Not so with nuclear warfare.

More than sixty years into the modern nuclear age, no one has figured out how to defeat a nuclear weapon. Even after thirty-five years of on-and-off effort, no one is confident that a ballistic missile can be consistently shot down.[40] Thus, almost all useful nuclear strategies with today's technologies involve deterrence. By contrast, while deterrence may have a role to play in information warfare, defense – notably passive defense[41] – is clearly appropriate and, in many contexts, dominant.

Several corollaries follow. The future of information systems security has far more to do with the future of information systems vulnerabilities than with information weapons.[42] By contrast, no one designs nuclear weapons against specific cities (even if some of them may be designed against particular types of sites such as underground bunkers). The details of a system's construction determine not only the susceptibility of a system to an information attack but what the effects of such an attack are. In nuclear warfare, the blast zone could be calculated with a fair degree of precision; the RAND Corporation developed calculator wheels that did just that. No such precision can be associated with an information attack, as Chapter 4 argues. Friendly fire and collateral damage are real risks of information warfare – all the more difficult for not being obvious at the time.

More fundamentally, nuclear warfare is real; Hiroshima was the blinding flash of the obvious. Information warfare is still largely theoretical.

---

[40] In August 2005, the head of the U.S. Missile Defense Agency reported that after spending $92.5 billion on missile defense, the United States had achieved a better-than-even chance of intercepting a single shot from North Korea. Patricia Parmalee, ed., "Industry Outlook," *Aviation Week and Space Technology*, August 1, 2005, p. 12.

[41] Israel withstood Iraq's Scud attacks in 1991 with one direct death (and sixteen indirect deaths). Iran suffered far more deaths from Scuds over the 1980–8 war with Iraq. The difference has little to do with the U.S.-supplied Patriot missiles with which Israel was armed, which mostly failed to destroy the missiles they were supposed to intercept. It probably had more to do with the superior construction of Israeli cities (mostly steel-reinforced concrete) and their lower population densities.

[42] So-called war dialers and code-breaking systems improve the efficiency of information attacks. Hacker tools reduce the need for typing and provide an easy-to-use interface. But neither changes the nature of information attacks. For more details, see Stuart McClure, Joel Schambray, and George Kurtz, *Hacking Exposed*, Berkeley (Osborne), 1999.

People have seen the detritus left behind by small-scale hacker attacks,[43] but no one has ever seen it work at the scale often claimed for it. Were such a demonstration to occur, observers may nevertheless plausibly, even if not reliably, tell themselves that disaster could not occur to *their* systems because their defenses are better. This ambiguity extends to specific attacks. A nuclear weapon leaves clear radiological traces.[44] If it comes by ballistic missile, its launch point can be calculated and its shooter identified. The effects of attacks in cyberspace are often hard to distinguish from mistake or accident. Without sophisticated computer forensics, it is difficult to say from where or from whom it came. Victimized institutions are often loath to give out any information at all. No greater contrast exists than between the clarity of a nuclear blast and the ambiguity of a cyber hit.

Early nuclear doctrine was driven by the reciprocal fear of a surprise attack. The Soviet Union had Barbarossa in its immediate past; the United States, Pearl Harbor. Both sides worried not only about force survivability (that is, "second strike"), but also how to detect indications and warnings of a "bolt from the blue." Is there an information warfare counterpart? One could try to collect evidence, typically from overseas, that an information attack were about to commence, but there is little confidence that anyone knows what to look for or could find it even if they did. Because the effects of information warfare are so specific to the target, an attacker usually has to scope it – not only the system's security architecture, but also how its owners use information– well before striking it. Supposedly, such probing will leave traces, but these days probing of every Internet-connected system takes place all the time. Thus, a probe has to be quite unique to say much about when an attack can be expected – and then only if the attacker plans an attack to follow closely behind the probe.

---

[43] Scale is in the eye of the beholder. Episodes such as Solar Sunrise (an orchestrated attack on unclassified DoD Internet sites in February 1998) and Moonlight Maze (an information attack that purportedly redirected some unclassified DoD printer files to Moscow) certainly excited people in the Pentagon, but the level of excitement greatly exceeded these incidents' measurable impact on national security.

[44] The smallest nuclear weapons had an explosive power comparable to twenty tons of TNT; no conventional weapon has ever come close to having that much explosive power.

Although both nuclear and information warfare are potentially complex, their complexity arises in different dimensions. The first-order results of a nuclear strike are straightforward.[45] But the nth-order results and meta-results of a nuclear confrontation are mind-boggling: if I do this, and he does that, and then I do thus, and so on. Because nuclear confrontations tend to push everything else into the background, they form a nearly closed system and confer an analytic credibility to nth-order interactions. The complexities of information warfare are broad because confrontations between attacker and host slide over a very complex information surface looking for holes and good places to erect barricades. But unless and until information attacks obviate all other issues, nth-order effects are more likely to wash out into other factors (such as legal issues, citizen–state relationships, and conventional military maneuvers) than to resonate cleanly through move and countermove. The overall outcome of a nuclear standoff may be more indeterminate in the long run (the prisoner's dilemma) than its cyber counterpart.[46]

The effects that information warfare can produce are limited in type, but that hardly proves they are limited in degree.[47] The results of a

---

[45] However, the aggregate effects of a full-scale nuclear exchange may well induce a drastic climate change, "nuclear winter."

[46] One factor that makes it hard to mount a sufficiently thorough insider attack on the large, heterogeneous, and variegated U.S. infrastructure is that it may involve subverting a large number of institutions' systems at once. The discovery that a serious subversion has taken place, especially if specific perpetrators or methods can be determined, introduces the risk that others will be on the lookout for telltale signs of an attack with a similar modus operandi. Finding a *second* such operation, in turn, further heightens the alert level and gives systems administrators something more specific to look for. Thus, the more operations have to be pulled off to achieve coercive effect, the larger the probability that the whole scheme will unravel before it has gotten very far.

An analogy may be made to the child's game of skunk. Players take turns rolling dice. If neither die is a snake-eye, then the player adds the total points shown on the dice to his or her total and gets the opportunity to roll again. If one die is a snake-eye then all the points from that turn are cancelled and the player's turn ends. If both dice show snake-eyes then the player score reverts to zero. Typically the game is played for 100 points. As the scoring requirement rises, the number of turns required before the first player achieves it rises much faster until it gets to the point where certain scores are nearly impossible to reach because players' score returns to zero so frequently.

[47] In late 2002, a full-scale attempt to cripple the Internet's central Domain Name Service (DNS) servers (the ones that play the major role in converting Web names to addresses and thence to paths) failed. According to Matrix NetSystems, the peak of the attack saw the average reachability for the entire DNS network dropped only to 94 percent from its normal levels near 100 percent. See Robert Lemos, "Attack on Net Servers Fails,"

full-scale attack on a nation's infrastructure depend, in large part, on the balance between man and machine in running the world. Whereas computers are everywhere, they are not equally important everywhere. Many parts of the world do just fine without them. Even in advanced societies, individuals can walk away from their computers. The services they support, while contributing to the quality of life, are generally not essential to its continuation. Living in a world without electricity, telecommunications, or even financial services would be quite wearing after a while – although, as the experience of even a modern city such as Sarajevo circa 1994 proves, such hardships would not, by themselves, be fatal to human survival. But many of the systems that make modern life livable, while computerized, still rely largely on humans for their logical processing, and still provide for manual intervention at various points. Attacking such systems would not necessarily damage or confuse them irretrievably.

Perhaps discussing information warfare as an activity on par with the very discipline that invented "overkill" is, in fact, overkill. To take it down a notch, can information warfare by nonstate actors be characterized as a form of "catastrophic terrorism"?[48]

2.3.1 IW Strategy and Terrorism

The belief that hostile computer network operations have a particular appeal to terrorists pervaded commentary in the wake of the World Trade Center attack.[49] Although the attackers had used computers and

http://news.com/2100-1001-963005.html, October 22, 2002; David McGuire and Brian Krebs, "Large-Scale Attack Cripples Internet Backbone," *Washington Post,* October 22, 2002, p. E05. Robert Lemos argues that the risk to infrastructure from hacking is exaggerated; even if penetrations take place, rarely are lives lost. Furthermore, those infrastructures that depend critically on continued accessibility to their computers tend to take security and resilience quite seriously. Robert Lemos, "Safety: Assessing the Infrastructure Risk," http://news.com.com/2009-1001-954780.html, August 26, 2002.

[48] Ashton Carter, John Deutch, and Philip Zelikow, "Catastrophic Terrorism: Tackling the New Danger," *Foreign Affairs,* November/December 1998, pp. 80–94. Although the article was directed against the threats of nuclear, biological, and chemical weapons, comparable attention was paid to cyberterrorism, with an entire page devoted to a proposed National Information Assurance Institute.

[49] For a recent survey of the cyberterrorist debate, see Gabriel Weimann, "Cyberterrorism: The Sum of All Fears?" *Studies in Conflict and Terrorism,* 28, 2005, pp. 129–49. This debate began even before September 11. Kevin Poulson, "Cyber Terror in the Air," Securityfocus.com, June 30, 2001. According to a recent study, 75 percent of Internet users

encryption to exchange messages, there is no evidence that they attacked computers as such.[50] Nevertheless, CNN correspondent Nic Robertson, when hearing about the first attack on the World Trade Center while covering a trial in Kabul, Afghanistan, "thought some kid had freaked out and hacked into some air traffic control or navigational database or something."[51] Even a few days later, the *Economist* argued, "The biggest nightmare is that Mr. bin Laden and his associates will acquire biological, chemical, or even nuclear weapons. . . . Another risk, says bin Laden-watcher Yonah Alexander, is that of devastating cyber-warfare."[52] Robert Kaplan, emphasizing the need for preemption against terrorism, also managed to portray cyber-attack as a step up from hijacking by observing, "Striking terrorist cells before they strike us – hitting not just hijackers but the computer command centers of our future adversaries before they can launch computer viruses on the United States, for instance – will need to be accomplished by surprise if it is to be effective."[53] As Roland Jacquard stated, "As the Sept. 11 catastrophe made tragically clear, the fanatic groups behind such terrorism will no longer content themselves with conventional, low-tech forms of attack. They aim to go further and hit harder by using biological and chemical weapons and no doubt bombs packed with radioactive material, not to mention cyberterrorism."[54] The British Intelligence Unit M2G argued:

In the not too distant future, there is a likelihood that command and control attacks, which blend cyber terrorism with physical terrorism, simultaneously seek

worldwide agree; they believe that "cyberterrorists" will "soon inflict massive casualties on innocent lives by attacking corporate and governmental computer networks." Five years before 9/11, the normally sober Walter Laqueur observed, "An unnamed U.S. intelligence official has boasted that with $1 billion and 20 capable hackers, he could shut down America. What he could achieve, a terrorist could too. . . . Why assassinate a politician or indiscriminately kill people when an attack on electronic switching will produce far more dramatic and lasting results?" "Postmodern Terrorism," *Foreign Affairs,* 75, 5 (September–October 1996): 24–36, p. 35.

[50] A *Washington Post* subtitle argued "On Guard against Cybercrime and Weapons of Mass Destruction, U.S. Is Blindsided by Attack at Home." Karen DeYoung, "Terrorism Warnings Focussed on Threat Overseas," September 12, 2001, p. A19.

[51] www.cnn.com, September 21, 2001, 3:46 P.M. EDT.

[52] *Economist,* September 22, 2001, p. 19.

[53] *Washington Post,* September 23, 2001, p. B5.

[54] "The Guidebook of Jihad," CNN, www.time.com/time/magazine/article/0,9171, 1101011029-180519,00.html, September 21, 2001.

to disrupt transport or telecommunications hubs; financial services or commerce; water or energy distribution; could also be manifest as hackers organize themselves more rigorously along the lines of criminally financed terrorist syndicates with specific ideological agendas and become more adept at social engineering to procure insider help locally.[55]

In late 2002, the U.S. consulting group IDC predicted that a major cyberterrorism event would occur in 2003, one that would disrupt the economy and bring the Internet to its knees for at least a day or two.[56] Offers David Tucker, "Whether they [barbarians] risk great battles or prefer innumerable small engagements, they will not hesitate to attack the American people directly. They will defeat us by hacking to death the information systems our economy and comfort depend on."[57] In the spring of 2005, the CIA conducted a war game to practice defending against an electronic assault on the same scale as the September 11 attacks and to test the ability of government and industry to respond to escalating Internet disruptions over many months.[58]

But it is too easy to get excited about cyberterrorists. As the *Washington Monthly* reported, stories that "discovered" that a twelve-year-old had come within a few keystrokes from unleashing a flood directed against suburbs of Phoenix, Arizona, were just not true.[59] The story, courtesy of Richard Clarke, who headed the National Security Council's cyberterrorism efforts, circulated around Washington, D.C., for years. Only when the "disaster" story got into the *Washington Post* did someone associated with the "imperiled" Roosevelt Dam discover the news item and correct it.[60]

In the years following September 11, it became more obvious that taking down the Internet is the *last* thing Islamic terrorists want to do; they need it too much. The U.S. invasion of Afghanistan, having destroyed

---

[55] Mailing list e-mail, November 11, 2002, 6:42 A.M.

[56] Ed Frauensheim, http://news.com.com/2100-1001-977780.html, December 12, 2002.

[57] David Tucker, "Fighting Barbarians," *Parameters*, Summer 1998, pp. 69–79.

[58] http://hcgtv.com/item/1025. The item added that the game "contravenes assurances by U.S. counterterrorism experts that such far-reaching effects from a cyber-attack are highly unlikely."

[59] Joshua Green, "The Myth of Cyberterrorism" *Washington Monthly*, November 2002, pp. 8–13.

[60] Barton Gellman, "Cyber-Attacks by Al Qaeda Feared," *Washington Post*, June 27, 2002, p. A1.

much of al Qaeda and driven its leaders into the mountains, has forced the movement to become much more distributed. Al Qaeda's leaders control few resources at this point. Instead, they have become an ideological nucleus for a more dispersed and, yes, networked movement. To act together in a way that seems remotely cohesive, such terrorists must be able to communicate. Having earned the opposition of almost every single government, they must do so without leaving footprints. The Internet is perfect for this.[61] Islamic terrorists, for their part, have thus come to rely heavily on it, sliding down its learning curve at a respectable clip.[62]

Correspondingly, terrorists use the Internet for three related purposes: recruiting adherents, distributing instructional materials, and exercising direct command and control.

Recruiting adherents is a multistep process. One step is channeling a vague resentment against the West into a belief in the rightness of violence against Western targets. Another step is converting this belief into a commitment to action and a set of connections that enable this commitment to be successfully exploited. For such purposes, the Internet has been described as a "vast recruiting ground – in effect, a new borderless Afghanistan."[63] These early steps are, by their nature, public and disproportionately phenomena of those parts of the world affluent enough to have routine Internet connectivity. The Web allows propagandists to create and float messages, test their effect, and refine accordingly – with, of course, shorter cycle times than erstwhile print services

---

[61] For more material on this topic, see, for instance, Nadya Labi, "Jihad 2.0," *Atlantic*, 297, 6, July/August 2006, pp. 102–8j; Paul Cruikshank and Mohanad Hage Ali, "Jihadist of Mass Destruction," *Washington Post*, June 11, 2006, p. B2; Audrey Kurth Cronin, "Cyber-Mobilization: The New Levee en Masse," *Parameters*, Summer 2006, pp. 77–87; Eshan Ahrari, "Al Qaeda and Cyberterrorism" www.atimes.com/atimes/printn.hmtl, August 18, 2004; Steve Coll and Susan B. Glasser, "Terrorists Turn to the Web as Base of Operations," *Washington Post*, August 7, 2005, p. A-1; Timothy Thomas, "Al Qaeda and the Internet: The Danger of 'Cyberplanning,'" *Parameters*, Spring 2003, pp. 112–23; Douglas Frantz, Josh Meter, and Richard B. Schmidt, "Cyberspace Gives al Qaeda Refuge," www.latimes.com/news/nationworld/world/la-fg/cyberterror15aug1,1,4439595.story, updated on www.jihadwatch.org/archives/002871.php, August 15, 2004.

[62] In 1997, Gabriel Wiemann, a professor at the University of Haifa, found twelve terrorist-related Web sites; by 2005 the count had surpassed 4,500. Coll and Glasser, op.cit.

[63] Labi, op. cit., p. 103.

permitted. Web sites that permit give and take allow potential recruits to picture themselves more easily as part of the movement than more passive many-to-one communications methods do. The suspects in the Toronto bombing case (arrested in early June 2006), for instance, were supposedly radicalized through terrorist sites and given direct advice by Younis Tsouli, a.k.a. "Irhabi 007," a Londoner accused of operating multiple al Qaeda–linked Web sites.

The instructional material distributed by Jihadists groups ranges widely: instructions on how to construct weapons and use cell phones to set them off,[64] operational and communications security tips, vulnerabilities in the public infrastructure, data on particular people to be used as targets, and lessons on kidnapping. Jihadists' Web sites also make it easy to raise money and distribute it in ways that make it hard to trace.

The Internet has been associated with the London Subway bombings of July 7, 2005, but the degree to which that association is relevant is hard to judge. *The Observer* reported, for instance that, "The official inquiry into the 7 July London bombings will say the attack was planned on a shoestring budget from information on the Internet," but although the Internet comes up several times in the report, it hardly plays a central role.[65] For instance, within hours of the blast, claims of responsibility were posted on the Internet in the name of al Qaeda,[66] but no independent evidence has surfaced to back up such a claim. Two of the bombers, Shehzad Tanweer and Germaine Lindsay, had purchased ancillary equipment for bomb-making (for example, face masks) over the Internet,[67] but not the primary materials. Little in the official report suggested that they depended on the Internet for their bomb-making "expertise." The

---

[64] "How to mix ricin poisoning, how to make a bomb from commercial chemicals, how to pose as a fisherman and sneak through Syria into Iraq, how to shoot at a U.S. soldier, and how to navigate by the stars while running through a night-shrouded desert." Coll and Glasser, op.cit.

[65] Mark Townsend, "Leak Reveals Official Story of London Bombings," *Observer*, observer.guardian.co.uk/uk_news/story/0,,1750139,00.html, April 9, 2006.

[66] Report of the Official Account of the Bombings in London on 7th July 2005, London, The Stationery Office, May 11, 2006, p. 8.

[67] Ibid., pp. 23, 25. By contrast, one of the seventeen alleged terrorists arrested in Canada in 2006 did use a library's Internet access to research bomb construction.

Internet was primarily used as a vehicle for propaganda, the report's analysis found:

As such, extremists are more and more making extensive use of the Internet. Websites are difficult to monitor and trace; they can be established anywhere and have global reach; they are anonymous, cheap and instantaneous; and it requires no special expertise to set up a website. The Internet is widely used for propaganda; training (including in weapons and explosives); to claim responsibility for attacks; and for grooming through chat rooms and elsewhere.[68]

Although direct command and control is entirely covert for obvious reasons, it still can be done over the Internet. The September 11 hijackers exchanged plans by signing up and logging into the same e-mail account, writing drafts of communications, then having others read it, and, if necessary, edit the draft as a form of reply – all without having to hit the send button. Such letters never circulated as e-mail.[69] They were not the last messages to circulate in this way. One of the men facing charges over the Madrid bombings, Hassan el Haski, allegedly set up free Hotmail and Yahoo! accounts as dead drops. Other groups reportedly disguised their coded communications as spam e-mails.[70]

Indirect command and control, however, can be quite overt. Speeches by leadership on the necessity or desirability of attacking certain classes of targets may be followed by actual attacks. Detailed instructions on kidnapping from an al Qaeda site in Saudi Arabia were followed three weeks later by the kidnapping of Paul Johnson, an aerospace engineer in Riyadh.[71]

Ironically, the two individuals least able to take advantage of the Internet directly (although they are no doubt aware of its multitudinous advantages) are Osama bin Laden and Ayman Zawahiri (al Qaeda's number one and two leaders) themselves. Apparently they produce no material

---

[68] Ibid., p. 31.

[69] This has given rise to the misperception that such missives never traveled through the Internet. They had to, of course, to get from Web e-mail servers (such as Hotmail's) to various clients. But never having traveled as e-mail, they used different and presumably less well-monitored ports.

[70] "Homegrown Terror: It's Not Over," *Maclean's*, June 19, 2006, pp. 19–26.

[71] Frantz et al., op. cit.

on the Web directly; it is always videotape to the media, and from there to cyberspace.

2.4 Conclusions

Hostile conquest in cyberspace is one aspect of information warfare, which, in turn, can be defined as the use of information to attack information. This definition fits computer network attack well because systems use information to manage, process, and protect information. But information warfare is about much more than information systems, although sometimes it is easy to think otherwise. It is fundamentally about information and the use to which information is put – making better decisions. So to understand the potential of hostile conquest requires understanding, on the one hand, the minute vagaries of the systems at issue, and on the other hand how decisions falter when information systems are not trustworthy.

For everyone, information systems ten, twenty, and fifty years hence will be far more pervasive. People everywhere will be more dependent not only on their functioning, but also on the soundness of the decisions they make. More nations will be wired, and residents of many states considered middle-income by today's standards may have information systems far more embedded in their day-to-day lives than is now true even in advanced computer-dependent regions, such as metropolitan San Francisco or Washington, D.C. The effects of information warfare may be more serious, the value of preserving information infrastructures will rise, and the leverage that their disruption may offer to potential foes will be that much greater. Cyberspace is undoubtedly becoming more consequential, but there should be no confusing it with nuclear war. That strategic it is not.

3

Information Warfare as Noise

Information at rest can be destroyed if its owners have failed to back it up. Information in motion can also be halted by attacks on communications system. There is yet one more way to destroy information, and that is by doing so indirectly. The bits, as it were, are untouched. But the credibility of the information is ruined by *adding* false information to it to the point where the victim must choose between misinformation (believing what is not true) or disinformation (being unable to believe what is true). Here, we argue, is the essential character of information warfare, something entirely consistent with the theory of information.

Information theory, for its part, was invented nearly sixty years ago by Claude Shannon[1] asking how much information can be extracted from a signal. In digital terms, if every bit is guaranteed to arrive error-free, then every bit can contain information. If some known percentage of bits is randomly flipped, the amount of reliable information in the bit-stream would be reduced. There are techniques (such as trellis encoding) that can, by adding bits, increase the confidence that one can recover the original signal, but they reduce the ratio of signal to bits transmitted, cannot restore 100 percent confidence, and require that the distribution of flipped bits be random and their expected error rate be more or less known. Randomly flipped bits can be seen as noise.

Noise, therefore, is the enemy of signal, and hence of the information carried by the signal.

---

[1] Claude Shannon, "A Mathematical Theory of Communication," *The Bell System Technical Journal*, July, October 1948, 27, pp. 379–423, 623–56.

Those who would wage war on information would, correspondingly, add noise to signals within information flows or information stores. Doing so would reduce the reliability of information that victims received, preventing them from making correct decisions, or at least from making them with sufficient confidence. They may not even make decisions at all. The more noise, the less signal and thus the less information flow – that is, the less correspondence between the perceived flow of information and the actual information that was meant to be conveyed. If noise levels are high enough, the data stream has no information in it.

A perspective that information warfare is essentially noise making seems to demote a profession. Don't information warriors attack by injecting alternative information into systems so as to control them and shape what their human targets know about the world? How can a theory meant to understand the effect of *ambient random* noise on information even begin to convey the effect of *deliberately altered,* and thus non-random, signal on the recipient's information flow? By what sleight of hand can something usually associated with war be better clothed in the metaphors of pollution?

To answer this question is to straddle the border between intent (to control) and effect (to confuse). In a world where the possibility of attacks is present, so is the doubt that what one knows is correct. From this follows an important distinction between not knowing something and knowing it wrongly. Information warfare, we argue, is best thought of as noise, certainly as it affects human decision making and increasingly as it affects systems decision making. One can defend against noise by redundancy and filtration, as the next section suggests. Yet one must first determine how noise-tolerant one's environment really is.

3.1 Disinformation and Misinformation

The distinction between disinformation and misinformation arises from different expectations. If one expects all information to be consistently true, then one will believe the occasional information that is false – misinformation. If, however, one has a correct sense of how much is likely to be true, then one will act more tentatively on all information, be it good or bad; this reduction in the usability of information is disinformation.

The trick is finding a balance between gullibility, and therefore susceptibility to deception, and cynicism, and therefore resistance to all information. Those who achieve a correct balance will still suffer from false positives – believing erroneous information to be correct. They will also suffer from false negatives – discarding good and true information in the belief that it may have been tainted. But striking a balance means they will not suffer from too much of either. And they can design alternatives as described in this chapter to increase the signal and decrease the noise. A balanced approach converts misinformation into disinformation. It does not eliminate the problem – or the advantages of giving such a problem to others. When Winston Churchill during World War I proposed dragging battleship silhouettes in the water, skeptics replied that the Germans would eventually realize they were being tricked; he responded that henceforth they would doubt their eyes whenever they saw any such silhouette, whether real or fake.

If information warfare leads not to doubt but excessive deception, the fault may lie in the victim's a prioris – expectations of a condition prior to its being validated. The basis for this difference lies in Bayesian logic – a way to convert evidence into judgment[2] and thus the cornerstone of

---

[2] Consider the following problem of Bayesian mathematics. Two urns both have three balls. In one urn, call it redmore, two balls are red and the other is green. In the other urn, call it greenmore, only one ball is red and the other two are green. Knowing this fact a priori (but being unable to differentiate the redmore urn from the greenmore urn based on inspection), you draw out the red ball from one of the urns. How likely, therefore, was it that you removed the ball from the redmore urn? Answer: two-thirds, the a posteriori probability. Why? Before pulling the ball out, there were six equally likely possibilities: that you pulled (1) red ball number 1 from the redmore urn, (2) red ball number 2 from the redmore urn, (3) the green ball from the redmore urn, (4) green ball number 1 from the greenmore urn (the other one), (5) green ball number 2 from the greenmore urn, or (6) the red ball from the greemore urn. Pulling out a red ball leaves one of only three equally likely possibilities: (1), (2), and (6). Of the three possibilities, two of them indicate that the redmore urn was chosen and one of them indicates that the greenmore urn was chosen – hence the two-thirds odds. To redefine the problem slightly, assume there is only one urn, with a 50:50 chance that it has more red balls than green balls. You pull out a red ball. The odds that this urn did, in fact, have two red balls and one green ball is the same: two-thirds. Now change the problem. Again, there is only one urn, but you are assured that there is a only a one-in-five chance (20:80) that it has two red balls and one green ball in it. You are told that these odds come from the fact that the one urn was randomly pulled from the back room, where there are four urns with two green balls and one urn with only one green ball. Here

modern statistical inference. Among its insights is that unless and until the volume of evidence is high, the conclusions one draws from evidence are strongly colored by the presumptions one makes going in.[3] If handed what your mother swears is a fair coin, it would take a very long run of successive tails before you would begin to doubt her. Until then, you would ascribe a run of tails or heads to luck. Conversely, if someone shady gives you the coin, you would need much less evidence before suspicions arose.

If you (as Hitler) are absolutely convinced that the Allies would invade Europe at Calais, then you are apt to absorb a great deal of contrary evidence, including the fact that troops have already landed in Normandy (it could be a feint), and still hold to old beliefs. If, conversely, you judged Normandy as a plausible landing point, even if Calais were more probable, then you would accord the evidence that favored Normandy its proper weight. Finally, if you knew that the Allies thought that you were strongly inclined to believe that Calais were the place, you would downgrade the evidence in favor of Calais as something that may quite possibly be manufactured – and so on down the hall of mirrors. A priori judgments affect how evidence is appreciated. Setting a correct a priori judgment is the first defense against information warfare because, by definition, it puts bounds on the excessive harm of accepting false information and the excessive harm of disregarding true information.

Most computer systems, however, are built with poor or at least rigid a priori logic. Once something gets past the hard filters (for example,

---

there are fifteen equally likely possibilities: (1) red ball number 1 from the redmore urn, (2) red ball number 2 from the redmore urn, (3) the green ball from the redmore urn, (4–7) green ball number 1 from a greenmore urn (remember, there are four such urns to begin with), (8–11) green ball number 2 from a greenmore urn (same four), or (12–15) a red ball from a greenmore urn. Pulling out a red ball validates one of only six equally likely possibilities: (1), (2), (12), (13), (14) and (15). Of the six possibilities, two of them indicate that the redmore urn was chosen and four of them indicate that the greenmore urn was chosen – hence the one-third odds but in the other direction. Even after the red ball was chosen, one would be justified in believing that it was more likely that one had a greenmore urn. So, the difference in a prioris turns what was odds-on conclusion that one had chosen a redmore urn to an odds-on conclusion in the opposite direction.

[3] The difficulty of ascertaining whether a long run of straight heads on coin flips indicated an unfair coin was famously mooted in Tom Stoppard's play *Rosencrantz and Guildenstern Are Dead.*

access controls or firewalls), it is generally held to be *legitimate* regardless of whether it is at all *reasonable* in nature.[4] Systems do not deal well with ground truth of the real world. Rarely are they asked to draw their own conclusions by sampling the world and comparing it to what common sense would label as normal. They are machines, and people like their machines to be obedient and deterministic. The a prioris of error are usually set to zero, and therefore false injected information becomes misinformation rather than disinformation. Although some electronic devices such as modems accommodate bit-level noise and integrated circuits can do internal error detection and correction, the recognition of logic-level noise is not as common apart from bounds checking in databases and related systems. Unless otherwise instructed, everything is signal – every instruction is valid, and every data item is processed as valid if it falls within specified bounds. Exceptions, such as the artificially intelligent programs employed by credit card companies to flag suspicious purchases, are notable for that fact.

It is unclear how long computers will remain so trusting or at least so binary – in that they accept information as reliable or reject it outright based on prespecified criteria. The emergence of intrusion detectors suggests that softer filters are coming that can sense some volume of noise, if not its nature.[5] That systems designers are coming to understand the ever-presence of noise in cyberspace is suggested by the emerging philosophy of protection through multiple barriers rather than – as is common today – placing hard barriers between insiders and outsiders and few barriers between insiders and system resources. But none of these protection

---

[4] A system may trust a message more depending on where it thinks it came from – which it can verify by means of digital signatures and certificate authorities (as well as reliable source pedigrees). But digital signature systems have to be reliable (who authenticates the authenticators?) and also require some capacity to judge that the trustworthiness of a message sender is greater than zero but less than one.

[5] Ironically, systems administrators complain that the difficulty of detecting and characterizing intrusions is frustrated by the fact that – relative to some imagined template of attack – information systems are inherently noisy environments. Noise, as such, does not announce itself as noise; if so, it would be filtered straightaway. What makes it noise is that it is carried as though it were signal when, in fact, it is only masquerading as such.

systems yet comes close to the sophistication of humans in discerning the real from the false.

Today's security problems, may, in retrospect, come to be understood as the shakedown process of taking a silicon creature from a cosseted, closed institutional environment and opening it to the real world. Naifs without street smarts are often rolled the first time out in the wrong neighborhood, and the Internet is proving to be a rough part of town. Fortunately, humans have not entirely abandoned their oversight over their systems. Much of what creates real information security is their constant monitoring of operations, looking out for any behavior that suggests that things are not quite right. Otherwise what should have been taken as noise will have to be a struggle to control a very fast but not particularly clever machine. But as growing sophistication shapes the balance between blind acceptance and blind rejection, information warfare can be more clearly seen as a phenomenon of noise.

3.2 Defenses against Noise

One can defend signal against noise by boosting the signal-to-noise ratio: strengthening the signal through redundancy, and weakening the noise by filtering.

3.2.1 Redundancy

Simple redundancy works against straightforward attacks. It entails repeating information in several places and transmitting it through several channels. Complex redundancy entails structuring information so that the various parts support the whole. In computer terms, it may be done by processes that save information in multiple encodings and locations. In human terms, it is the bundling of primary information together with supporting information and consequences. Complex redundancy takes advantage of the fact that information often presents itself as a linked agglomeration of a few important facts and a larger number of facts of great but not overwhelming import all interspersed among facts of individually minor import.

Human language is usefully redundant. If driblets of white-out are splattered on a piece of paper or if a slice of a page is lost in duplication, most of its meaning can be reconstructed. The capability arises from how words and sentences are built, coupled with the normal expected patterns of word use, the expectations of consistent context, and the layering of information throughout a paragraph. A sentence commonly seen on matchbook covers argues that if you "cn rd ths thn y cn gt vry gd jb" and proves that most vowels are redundant. But there are exceptions; mangling proper names or substituting "now" for "not" will change information in important ways. English is redundant but not systematically so.

A more systematic approach to protecting information is to structure it purposely so as to permit recovery in the face of noise. Trellis encoding is one such scheme that works at the bit level if noise levels are relatively low and random. Human memory embodies such a principle; most information can be remembered even if random neurons die. A mathematical analogy of a neuron system is the Hopfield net, which stores information as distributed relationships among nodes in such a way that the destruction of specific connections results not in lost memory, but in less certainty and less capacity for further memory. In Fourier transforms, unnatural and thus high-information waveforms such as peaks, steps, and serrated edges can be expressed as a converging infinite series of smooth sine waves. A good court case is one in which the critical piece of the argument – guilty or not guilty – can survive the destruction of any one or two pieces of evidence. The totality of evidence is what counts and thus small amounts of noise can be introduced into it without effect. Again, this is not the way that computers, for the most part, are commanded and controlled, but it could be and one day may be.

How much noise can an information conduit withstand before its value is totally destroyed? There is a good deal of information worth acting on, even if it is unlikely to be true, because its availability narrows an otherwise vast search space. Take encryption; the 128-bit Advanced Encryption Standard generates messages that would take ten trillion trillion trillion keys to crack. A key that has only a one-in-a-thousand chance of being correct is clearly worth trying if the message matters. A set of such 1:1000 keys that has been bloated by the addition of ten times as many bogus keys is still quite valuable. Conversely, there is information worth acting on only if the likelihood of its being true is very high and thus is very sensitive to noise. People in rich countries can afford to be leery about ingesting harmful substances; they place a much higher value in avoiding food that has been falsely portrayed as safe (false positives) even if their caution causes them to reject food that has been falsely portrayed as suspicious (false negatives). Even a feeble rumor that a particular potion is poisonous or that a particular food is carcinogenic is enough noise to ruin its sales. It is easy to sully the reputation of a stranger and hard for someone so affected by noise to recover it – even if someone with a good reputation can withstand attack more easily.

3.2.2 Filtration

The effects of noise can also be mitigated by filtering it out. In making or using a filter, it helps if the receiver knows something about the source that allows it to differentiate signal from noise. It helps more if the adversary is ignorant of the correct source – but it is more important that the adversary cannot duplicate what the source knows so that it cannot deceive the filter.

This is nicely demonstrated in electronic warfare. A receiver can get around jamming by knowing the location of the transmitter and focusing on signals coming from that source to the exclusion of anything coming in from a side angle. The signal is likely to get through if the adversary does not know where the signal is coming from or knows but cannot place a transmitter in between it and the receiver. Similarly, a receiver can listen for a signal coming in on a unique frequency and filter out all the noise radiating on other frequencies. Refinements of these defensive techniques include pulse broadcasting (concentrating energy into a short time interval) as well as frequency hopping (if the enemy can detect and duplicate the transmitter's frequency) or other ways to encode a signal through a broad bandwidth (such as code-division multiple access, or ultra-wideband broadcasting). Error-correction techniques permit partially obliterated messages to be reconstructed. One may also be able to exploit known redundancy or other characteristic patterns in order to extract a signal from a noisy environment.

Filtration is also behind many principles of computer network defense. A password is a filtration device by virtue of separating legitimate from illegitimate access attempts.[6] An automated teller machine (ATM) that fails to recognize anything but legitimate keystroke combinations is another example. Some filters work in reverse, attempting not to select legitimate traffic from a total flow, but to filter out illegitimate traffic based on certain characteristics. Antivirus programs and intrusion detectors filter bad information for inspection and can thereby help reveal uniquely suspicious behavior patterns. Countering information warfare depends, in part, on knowing how much is coming at a system and how much gets through – so as to accord due weight to the received signal and to prepare additional measures (such as an otherwise redundant channel) in order to improve the signal-to-noise ratio cost-effectively.

Humans also use a sophisticated and mostly sensible variety of filtration techniques. We credit certain people as being more reliable or sensible than others. We understand that testimony from a person with a financial or emotional interest in some area is less credible than testimony from a disinterested party. We pay attention to someone's emotional state in evaluating what he or she says. We also filter by context: news considered a priori unlikely (for example, that an obscure candidate is suddenly winning an election) is held at arm's length until we receive confirming information. We also infer causality based on such principles. Without better evidence, we are more likely to infer that sudden darkness is caused by the passing of a large cloud than it is by a solar eclipse. Doctors are taught that when they hear the clatter of hooves, first think horses not zebras. But people also filter out information that does not accord to their preconceptions, violates their prejudices, deflates their ego, or hammers away at their comfortable expectations. And we take as credible images from media that have yet to seriously lie to us – those who would use video morphing to show world leaders in uncharacteristically embarrassing acts know that this trick will not work a second time. Thanks to Hollywood having gotten there earlier, we may not even fall for the trick

---

[6] The use of public-private keys as access devices has the advantage that an adversary knowing the mechanism by which a message is authenticated is not tantamount to the adversary's being able to duplicate such a message.

the first time. Those who would practice deception exploit the tendency of people to set their filters badly or at least uncritically.

Because of the increasingly complex demands being made on and largely accommodated by information systems, the filtration surface of the computer must be complex. It hardly suffices to make a once-and-for-all determination of who gets in and who does not. Our desires are more expansive than that; for example, we want the ability to add or delete authorized users remotely, processes that act as though they were users, and multirole permissions. So, security becomes harder. The fancier an operating system (for example, Windows 2000 with its nearly 50 million lines of code), the more places bugs can lurk. Web plug-ins (many developed after the operating system appeared) make configuration control a potential nightmare. Without imposed controls, no one can predict in advance which network-aware applications users have running on their desks, how such applications interact with what was originally installed, or how applications run by one user interact with those run by another. A filtration regime that exhibits all of its complexity on its topmost layer is bound to have flaws noticed only by the wrong people. Those whose complexity is distributed in layers tend to be more robust.

3.3 What Tolerance for Noise?

Noise, ever-present in the physical environment,[7] is also ever-present in cyberspace. Some noise is ambient; it is an artifact of human and computer randomness. Some noise comes from error and bad design. Only a fraction is deliberately induced. It is often hard to distinguish one source of noise from another – but not always necessary: filtering methods used in one situation are helpful in another. Professionals understand that many techniques that guard systems against deliberate attack are also useful against accident and error. They include provability technologies to determine whether code does what it should, network management as a professional discipline, and safety engineering that ensures that even bad inputs do not create hazards.

---

[7] Thermally speaking, it exists whenever temperatures are not at absolute zero – and even empty space is filled with quantum noise.

Likewise, the relative ability or inability to thrive among noise is a feature of systems. Some systems are noise-tolerant, others noise-intolerant. Knowing how tolerant is a good starting place in knowing what defenses to erect.

3.3.1 Tolerance in Real Environments

With modern societies descending deeper into the swirl of cacophonous information, it might seem their tolerance for noise would also be increasing.

Well, yes and no.

The industrial revolution, with its increasing emphasis on precision, in fact, has led to far *less* tolerance of noise, if noise refers to random variation. Random variation is the difference between the measured and the actual characteristics of an object and by extension between information and reality. Eli Whitney's invention of interchangeable parts in lieu of hand-fit components required that the error between the part itself and its design be sufficiently small. This called for advanced techniques of measurement and control – a test that Japan has famously passed. The country improved its manufacturing art through Toyota's development of synchronized parts delivery (just-in-time), which admitted of little deviation, and a quality-control regime, the Toguchi method, based on minimizing the total variation across the entire manufacturing process. A modern chip manufacturing facility is a highly controlled environment by dint of being a well-filtered one; one speck of lint can destroy a silicon chip. A bad line of code can disable complex software. Bad data can reduce the credibility of a complex calculation.

Outside the factory gates, it is growing harder to find a piece of public terrain that is not slathered with advertisements, or a minute on the airwaves in which one or another commercial product has not been insinuated into the stream of words (for example, the Washington Redskins recently moved to a new stadium named Fedex Field, ensuring mention of the delivery company's name in innumerable sports broadcasts). The world has gotten noisier because it has become more incessantly competitive – especially in the United States. We live in the attention economy because time is the irreducible constraint on human input. Many of the new communications technologies such as cell phones and e-mail do a good job of extending the workday into the home. Customers demand them because of the impetus to know just a little more than one's competitor. But the quality of this extra knowledge is low. Such is an environment with a high noise tolerance.

The value of noise-free information can vary greatly. Assume someone were offered a dozen bits of absolute reliability. A greedy person could do far worse than to ask for the name of the stock guaranteed to offer the best five-year return; twelve bits suffice to indicate its unique position on the *Wall Street Journal's* "NYSE" page. If the information were absolutely true, one would borrow to the hilt to buy shares of its stock. So is Wall Street noise-intolerant? Quite the reverse; what moves the Street on a day-to-day basis is rumor, proto-news, and someone's best guess. Analysts must collect a large amount of information and analyze it without bias if they are to outperform the indexes. To a seasoned analyst, the addition of more noise is unlikely to make much difference – the ore from which fortunes are made is low-grade to begin with.

How the body is defended against disease is quite sensitive to which parts of the body are more or less noise-sensitive. In theory, the immune system uses a binary decision mechanism: if it encounters a protein originally present in the embryo – one that is thus natural to the human body – let it pass. If not, destroy it. In practice, the human immune system is approximate and heavily layered; barriers and intrusion detection play large roles. The immunoglobulin molecules of white blood cells cannot read proteins precisely: there are matches, misses, and in-betweens. There is always the risk of responding too weakly to beat off infection and responding too strongly and thereby stimulating allergic or autoimmune reactions. Fortunately, the human body is analog. Humans can lose a certain fraction of almost all organic functions and not suffer from or even realize it, especially if the organ repairs itself over time. Three parts of the human body, however, are essentially digital, in that the molecular equivalent of bit errors can lead to noticeable problems: brains, eyes, and germinal cells. There, an entirely different immune system takes over, one more dependent on barriers and ready to kill off entire cells at the least sign of trouble. Why the difference? The human body, overall, is a noisy, and thus a noise-tolerant, environment and has no alternative but

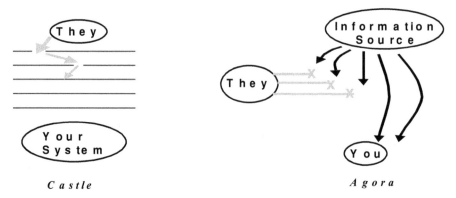

Figure 3. Protecting Castles and Agoras

to apply rough-and-ready filters to keep itself alive. But noise-intolerant organs use altogether different approaches (for example, the blood-brain barrier) against the microbiological equivalent of information warfare.

3.3.2 Castles and Agoras

Similarly, one can characterize information systems along a continuum anchored by two ideal types: castles and agoras. Castles protect noise-intolerant environments; agoras are noise-tolerant, indeed noisy, environments. Castles characterize a nation's critical infrastructures – military C^{4}ISR systems, funds transfer, safety regulation, power plants and similar industrial facilities, telecommunications switching systems, as well as energy and transportation control points. They are generally self-contained units, access to which can be, and ought to be, restricted. Agoras are the great consumer and political marketplaces of cyberspace in which the reduced security against malefactors (and susceptibility to random disease) is the price paid for the dense interactions and potential learning experiences that contact with strangers permits. It is as hazardous to use the rules of the agora to govern the castle as it is constricting to enforce the castle's norms on the agora.

Figure 3 illustrates a difference between strategies for castles and those for agoras (note the analogy to Figure 1). The rogue's route into the castle must surmount the horizontal lines on the left. To enter, the foe must breach several walls. Each wall has several potential holes that are hard for each side to see; both sides spend resources looking for them. One hole suffices to get adversaries through any one layer; this is an inherent advantage for the offense. Yet, in a multilayer system, news (for example, from intrusion detection devices) that the adversary has gotten in may prompt a rush of defenders to bolster the second wall (such as through tightening up permissions that users enjoy) as well as to find and plug the original hole. This process may be repeated through as many layers as the castle has. Total failure permits foes to raise havoc inside the castle's walls.

By contrast, the vertical lines on the right depict the contest in the agora. Along these lines flow information from the environment to the decision maker. The attacker is trying to induce noise along the path and thereby deceive or paralyze decision making. The defender's strategy is to open up as many paths as necessary to negate the effect of the noise. Multiple paths raise the likelihood that one or another path is sufficiently noise-free. If they are all noisy, then multiple paths allow the defender to ignore whatever signals do not show up in more than one path.

Ironically, the strategy of redundancy that works well in defining agoras and physical systems is counterproductive in defending castles. Creating redundant paths to the same end works against the defender, since it forces administrators to secure more access points and raises the odds that one of them has an exploitable flaw.

The proper mix of redundancy and filters should reflect the noise-tolerance of the environment to be protected. It may be less profitable to suppress attacks on information by filtering out ambient noise because there is so much of it already present in agoras. Similarly it is almost impossible to defend castles by raising the flow of information from the victim to the information system (except insofar as the victim vets questionable commands in order to filter out the dangerous ones).

A more general example is a computer network with servers and clients. Servers are few. They hold the corporate data files, are run by systems administrators for purely business reasons, may be physically secured, and may use a proprietary or tightly secured operating system with most of the troubling features disabled. Clients are many. They access corporate data files; if corrupted, their data may be refreshed from servers.

Many in number, they host multiple applications; some contain down-loaded games or links to nonbusiness Web sites. They often travel, use commercial operating systems with very little functionality disabled, and are owned by those worried about things other than security. One way to secure a client-server system against corruption may be to guard the servers and look askance at the clients. Sensitive information would be made inaccessible to clients except under stringent conditions. The valid-ity of information entered by clients would be considered suspect until verified. Some critical files of the servers might be placed on a medium that could not be physically rewritten to (for example, a read-only chip or a CD-ROM). Techniques would be used to prevent infected clients from flooding servers, thereby denying service to others. The list of legal messages that could pass from clients to servers would be watched closely to ensure that no message or combination of messages could cause great harm to the servers. An agora such as a product catalog may have to be linked to a more sensitive castle such as a database that holds credit-card records; between the two would sit sophisticated and very picky filters.

Complex networks may, by housing both castles and agoras, be fur-ther likened to airports.[8] Airports have parking lots, ticket counters, and shops where anyone can go; passenger areas accessible only by ticket-holders who have cleared security; and work areas limited to authorized employees and guarded by biometric systems. Corporate Web sites have security rules different from those of internal intranets. Product engi-neering shops may have the tightest rules of all; many in Silicon Valley are air-gapped.

3.3.3 Hopping from Agoras to Castles?

Can the castles of complex systems be attacked through their agoras? Among the military's networks is SIPRnet, a supposedly air-gapped sys-tem bolstered by cryptography, and the NIPRnet – essentially the ".mil" domain of the Internet. The latter is protected, on average, no worse but not spectacularly better than, say, a corporation would protect its own dot-com domain. SIPRnet has supposedly never been penetrated from

---

[8] See the survey "Securing the Cloud," *Economist*, October 24, 2002, p. 17.

the outside, but viruses (notably the I–Love-You virus) have been known to get past its barriers somehow.[9]

With any network operating at multiple levels of security, one must wonder how well access to the normal networks can influence what normally goes on via the more secret networks. Here, the three layers noted in Chapter 1 – the physical, the syntactic, and the semantic – come into play.

Perhaps the two networks are not as physically independent as they appear; for example, suppose that they use similar routers. Faults introduced into the normal net may create hardware conditions on common facilities that affect the secret net. Attacking an installation's computer-controlled air conditioning may raise temperatures enough to force the secret net's servers to turn themselves off. Faults introduced into the normal net may force people to use the secret net more intensively, thereby creating congestion and other problems. Such problems may be direct (such as unanticipated traffic levels) or indirect (for example, intrusion detection systems that react to traffic patterns pushed off the norm because of unanticipated usage).

Perhaps, too, the two networks are not so syntactically independent. They may use similar hardware and rely on poorly engineered software partitions to keep traffic separate. Attacks on the normal net can weaken the barriers directly, or indirectly (by inducing systems administrators to set parameters badly, for example). Or, while the two networks are electronically separate, media infected while on the normal net then infect the secret net upon transfer (because security rules are designed to restrict the flow of information, media that touch the secret network are therefore embargoed and thus no longer allowed to touch the normal network). The connection may be human; for example, perhaps someone with two terminals gets confused over which is which and puts sensitive material meant for the secret net on the normal net. Errors introduced into utilities (such as PKI-servers) used by the normal net may be transferred by hand to similar utilities used by the secret net.

Finally, are the two networks sufficiently independent semantically? Many people are on both networks. More commonly, people on the secret

---

[9] Charles C. Mann, "The Mole in the Machine," *New York Times Sunday Magazine*, July 25, 1999, pp. 32–5.

net communicate with and, to that extent, are influenced by people on the normal net. Even if the normal net is limited to garrisoned and support forces while the secret net is reserved for fielded forces, relationships among U.S. forces span normal/secret net lines; indeed, they may span lines between the secret net and the broader Internet. Witness the speed with which detailed accounts of pilot Scott Grady's ordeal (after having been shot down in Yugoslavia) spread by e-mail throughout the entire military and into the press. Conversely, misleading or simply low-value information fed into the normal net can, by word of mouth, affect people who, in theory, get their information only from the secret net. Ironically, the more secure the normal net is considered, the more likely that someone purporting to be a legitimate user will be believed as such, and thus the more influential such a person would be if he or she were to show up on the secret net.

Whereas the spread of influence may be from the normal net to the secret net, the spread of telltale information often runs the reverse course. Preparations for a campaign for which the details are restricted to secret net sites may, nevertheless, leave traces on normal net sites. Campaign plans devised in great secret may create logistics and deployment requirements that can be read from tapping normal traffic, which carries logistics information. Even if operational security is imposed to squelch such information, the fact that people are being moved from the garrison to the field may be revealed by something as elementary as an unexplained fall-off in e-mail from those redeployed. The imposition of operational security may be detectable from otherwise unexplained changes in e-mail, shifts in the patterns of Web access, and message traffic reduced in volume or sensitivity.

3.3.4 Castling Foes

One basis for confidence in the ability to protect cyberspace, especially as a castle, is that it remains rooted in the material world. The physical possession of the various elements of cyberspace allows the owners to use many, albeit somewhat blunt, tools of control. Systems administrators can cut off infrastructures electronically from the rest of the world by finding and then disconnecting the wires, and monitoring and then shutting off or perhaps jamming electromagnetic emissions. They can keep operating systems on media that cannot be written to. They can authenticate users to the system based on their possession of tokens (such as keys) that are not supposed to be distributed to the other side. Machinery controlled by corruptible computers can be designed for manual override. Primitive as such techniques are, they do permit systems administrators to reduce the degree and depth of damage if everyday security mechanisms fail.

By so doing, administrators can tighten cyberspace, under stress, into a closed little ball made impervious to the outside world.[10] They rarely do so, however, because the benefits of isolation rarely exceed both the costs of tightening up plus the pain felt by users as they are deprived of services and access to the outside.

In war, circling the wagons in cyberspace may be justified, or at least justified by a security community that arrogates to itself key access decisions when the risk to an information system and its contents becomes great enough. One thereby recovers confidence in the information acquired from inside but only by limiting the input of a much larger store of low-grade information from outside. Low-grade information may not be missed anyway. If such information is routinely corrupted, the cost of collecting and filtering it may begin to exceed its modest value. Under stress, the desire to separate insiders and outsiders rises. Such habits are strengthened by the tendency of militaries to be closed and well-defined entities. Membership in formal militaries is binary; one is in uniform or one is not. Those inside are accorded rights and privileges denied to those outside. There is a high degree of mutual loyalty, born, in part, of the historical expectation of being placed in common danger with one's peers.[11]

---

[10] Not every process can be so secured – for instance RF transmissions have to travel over a common RF medium, but the ability to disrupt RF signals at the physical level by no means guarantees the ability to corrupt them at the syntactic or semantic level.

[11] This is particularly true of the FBI, if the Webster Report is any indication. U.S. Department of Justice, *A Review of FBI Security Programs*, March 31, 2002. The FBI's case files migrate into the Automated Case System (ACS), where most are accessible to every agent of the FBI and to no one else (the FBI runs an air-gapped system). The Robert Hanssen case shows that confidence in agents may be occasionally misplaced. As Oliver Revell's account suggests, the FBI has a great many similarities to the Marine Corps in

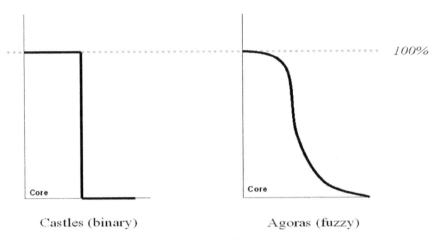

Figure 4. Degrees of Membership in Closed and Open Organizations

This binary approach is depicted in Figure 4, which arrays individuals from left to right by their degree of membership within an organization. Someone of degree 1 is indisputably a member of the organization, and the bulk of his or her relationships are within the organization; someone of degree 0 has no relationship to the organization whatsoever. The military may have values at or near 1 for people on the inside of the organization, and 0 or close to it for those on the outside.

Civilian organizations, by contrast, are more permeable and open.[12] Mutual loyalty is not as strong. People on the inside can often be fired at will; their jobs can be outsourced. Correspondingly, people on the outside

making inside/outside distinctions. Oliver Revell, *A G-Man's Journal*, New York (Simon and Schuster), 1998).

[12] Such organizations are becoming increasingly open. Rob Cross, Stephen Borgatti, and Andrew Parker, "Making Invisible Work Visible: Using Social Network Analysis to Support Strategic Collaboration," *California Management Review*, Winter 2002, p. 25. Over the past decade, significant restructuring efforts have resulted in organizations with fewer hierarchical levels and more permeable internal and external boundaries. *Economist*, "*Economist* Survey: The New Organization: Partners in Wealth," January 21, 2006. In the traditional organizational structure, units were either within the organization and, as Mercer Delta's David Nadler puts it, "densely connected," or they were outside the organization and not connected at all (ibid., p. 16). Transactions with external suppliers were at arm's length. By contrast, companies today cohabit with a vast number of joint ventures and strategic alliances, some more and some less connected. The line between what is inside the corporation and what is outside the corporation, once so clear, has become blurred.

can have various relationships with the organization as adjuncts, lecturers, vendors, service suppliers, collaborators, and so on.[13] An analogous curve might score people from (notionally) 1/2 and 1 if on the inside and 0 and 1/2 if on the outside, but some outsiders will have greater access to information within the organization than some insiders. Membership is a fuzzy set.

Given a choice between reaching outside of itself and losing internal cohesion, military instinct favors cohesion; civilian instinct favors extension. Military organizations put great stock on hierarchy and command and control. Civilian organizations are more likely to use networked forms, matrix management, and a complex set of formal and informal incentives. They are also more willing to embrace coalitions and consortia to get work done.

In cyberspace, the more something is closed, the more impervious and thus putatively safer it is. DoD's overall tendency as an organization is to be wary of openness. In Bosnia, it was the U.S. Army, in contrast to North Atlantic Treaty Organization (NATO) partners, that ventured out in full battle dress and four-vehicle convoys and spent little time or effort relating to the population. A similar contrast between U.S. and U.K. forces has also been observed in Iraq.[14]

Trends in today's national security and commercial environment alike, however, may force DoD's attention to the agora. Binary membership and hierarchical organizations are falling out of favor everywhere – even in the DoD. Why? First, in the information era, DoD needs to tap into smart people, many of whom do not work for DoD directly. More generally, DoD needs access to smart capabilities of all sorts that it does not own and could not own (for example, people who are valuable precisely because they have a breadth of experience outside DoD). Second, DoD needs to attract, retain, and get the most out of coalitions with other militaries. Third, it is probably better to work with rather than against the people on whose behalf you are fighting. If they are at all advanced, then getting

---

[13] The RAND Corporation, for instance has as many adjunct employees as full-time employees at any one time.

[14] *Economist,* "How to Do Better," economist.com/world/story_id=5300181, December 14, 2005.

them to contribute to your knowledge is increasingly a matter of melding what they know with your systems and, in some cases, linking to the systems by which *they* store, refine, and circulate such knowledge.

In light of such trends, information warriors may better use their art not so much to control an adversary's information system but to induce the adversary to close itself off. It is akin to raising a din in hopes that others would huddle under a cone of silence to filter out the noise and thus become deaf to the world. True, one can have a full and free exchange of information with those in the outside world without their getting into the computer system. If the information is limited to the exchange of carefully scrutinized news, then it can be thrown over the transom, so to speak (e-mailed to a firewalled Web site or shuttled to discontinuously connected systems, for example) with little danger that recipients can find their way back into DoD's systems. This still leaves the risk that bad information delivered through reliable (that is, nondistorting) channels is fed into and thereby corrupts decisions. In that case, information warfare at the semantic layer is akin to everyday classic deception.

As long as information can be passed without letting someone else into one's systems, it is possible to roll cyberspace into a ball and still exchange information. But for how long? Technologies that transfer information through interaction rather than simple message passing may become the norm for sophisticated information. Examples include (1) multimedia collaboration with video-teleconferencing, white-boarding (the ability of both sides to view and mark up a commonly accessible document), and real-time access to knowledge bases; (2) software agents that can sustain interactions with other people or their software agents without the owner's sustained attention to every single interaction; and (3) shared simulated environments where people or agents interact within the confines of a shared context (as in simulated combat). Today's rudimentary forms show why the ability to share information with others may come to require connecting information systems. Whether it is done by allowing others into one's own enterprise infrastructure or by creating a more carefully shared space is secondary.

We end with a metaphor. Trees are systems that start with a trunk, and work their way out as limbs, branches, and sprigs, respectively. A living tree even without its leaves can be characterized by detail; it maximizes surface area per bulk. To describe the tree in information terms take a lot of bytes. Trees, however, show their health from the outside in – from bare sprigs, to branches, limbs, and, finally, the trunk. A dying tree has less detail, less surface area, and, ultimately less information. Thus, an important secondary purpose of attacking an adversary's cyberspace is to get it to react by reducing the flow of information across its membrane, cutting it off from specialized differentiated knowledge, and, in effect, minimizing its exposed surface area, shedding leaves, sprigs, branches, and limbs as it does so.

Those who can keep their enemies in the castle own the agora.

3.4 Concluding Observations

Information warriors exist to introduce noise into an adversary's information system. They may want control, and sometimes they may get it, but there are reasons to believe that such goals are hard to attain today and getting no easier. Focusing on noise as the consequence of information warfare reveals redundancy and filtration to be critical features of information integrity. Noise tolerance becomes an essential distinguishing characteristic among systems, which can thereby be arrayed on the continuum between noise-intolerant castles (such as military C^4I systems) and noise-tolerant agoras (such as chat rooms).

Noise has its place in the arsenal. It can help paralyze or at least inhibit command and control. Under attack, decision makers are not sure that what they know is true, and neither are those who are supposed to carry out their decisions; so, commanders cannot exercise control over their forces or instruments. Even if all a prioris are accurate, the victim's sense that all information flows are fraught with error makes almost every resulting decision tentative. Nothing can be done boldly, cleanly, or decisively. Fall-backs must always be in place. Even good information cannot inform decisions (for example, to give or take orders) as well as it might, and its value is correspondingly reduced. If nothing else, information warfare raises the amount of filtering that must be done to use information systems confidently. Both data and processes must bear greater scrutiny and scrubbing before they can be used. Relationships based on trust among strangers (who must determine, for example, whether the client is here for a legitimate purpose or to loiter and then vandalize, or whether the merchant's goods are reliable) require higher levels of validation. Thus transaction costs might rise if legitimate colleagues, clients, or suppliers are not to be rejected. When what you hear contradicts what you know, then even memory (yesterday's information flows and lessons learned from them) has to be scrutinized once again.

Nevertheless, making noise is not exercising control and it hardly comports to anyone's idea of conquest. Nor, on that basis, can electrons be labeled the "ultimate precision weapon." Ascribing precision to noise violates intuition and correctly so. Indeed, to the extent that noise destroys information, or, similarly, destroys the credibility of information, then it is entropy and thus quite imprecise. Many advocates of information warfare, especially overseas, refer to viruses and worms as offensive weapons. Their release into the wild provides a good deal of noise, particularly in the media, but can any weapon be *less* precise? One knows neither its target population nor its effect on them. Hence information warfare is more likely to wreak confusion than destruction.

4

Can Information Warfare Be Strategic?

Computer hacking exists and its effects are not always trivial. Writers of viruses and worms can also annoy a great number of people. Authors of similar malware have succeeded in collectively turning millions of computers into zombies, programmed to spam Web sites or mailboxes on command.[1] Hackers have stolen such personal information as credit-card or social security numbers from feckless owners of databases.[2] Intelligence agencies have registered a fair amount of success trolling in files and pulling out interesting albeit generally random pieces of information. Indeed, one can state with a great deal of confidence that any information system connected to the Internet that gives access to enough trusted people, such as employees, can be tricked into giving the same access to a hacker; it only takes one sloppy user to open the door inadvertently. As long as hackers are not terribly fussy about who they annoy, whose machine they take over, or whose identity they acquire, they can always find some weak links out there.

All this, though, remains a far cry from getting a specific system to do what you want it to do, which is the sine qua non of using information warfare for strategic effect. Granted, the ability to exacerbate

---

[1] In October 2005, Dutch police arrested three people who created a botnet with 1.5 million computers – or perhaps many more – and used it to extort money from U.S. companies. Bruce Schneier, www.schneier.com/blog/archives/2005/12, December 22, 2005. See also Brian Krebs, "Invasion of the Computer Snatchers," *Washington Post Sunday Magazine*, February 19, 2006, pp. 10–15, 23–9.

[2] For example, see Fred Katayama, "Hacker Accesses 2.2 Million Credit Cards," www.cnn.com/2003/tech/02/17/creditcard.hack/index.html, February 17, 2003.

the usual chaos in an adversary's economy or military can be advanta-
geous only if used sparingly and wisely. But if that were all one could
expect from information warfare, it would hardly rate the attention it
has received both in the press and in the Pentagon. The real value comes
from being able to take control of someone else's systems to the point
where they yield their secrets, crash, or, most insidiously, misguide those
who use them. These systems must be more than just random ones. To
justify the attention paid to information warfare, the systems that can be
struck must be prominent among the systems on which the victim itself
relies.

In other words, there is a great difference in usefulness between a
technique capable of culling those who fall out of the security flock and
one that can target the alpha males, the strategic heights of opponents'
information systems. How well might information warfare work against
an adversary who knows the attackers are coming – and indeed are coming
all the time?

This chapter does not answer such a question directly, nor could it. As
the discussion that follows should make clear, forecasting in the particular is hard enough; forecasting in the abstract is impossible. Offensive
information warfare capabilities are highly classified, and for good reason. The quality of an information warfare attack generally comes from
its ability to deceive defenses. If the precise techniques were known, then
the precise defense against such techniques can usually be engineered
straightforwardly. Rarely, for instance, are vulnerabilities discovered and
publicized without patches being invented within days after the faults are
made public. For similar reasons, the types of defenses erected against
attack are almost never revealed except at the most general level (for example, the installation of firewalls or antivirus systems). Attackers with such
specific knowledge would have an easier job. At the very least, they could
avoid certain attack methods; if lucky, they may correctly infer what holes
remain by analyzing what is *not* listed among defenses.

What this chapter hopes to illustrate, in a general way, is the difficulty of
knowing in advance that a system can be targeted and exploited usefully –
two different tasks. Getting into a system may be hard enough. Taking and
maintaining control over it against an intelligent defender is, as illustrated,
quite something else entirely. Predicting success before the defender is

engaged is close to a fool's errand. That noted, as reported several years ago:

How the Defense Department will build a target folder for information warfare is now being debated and tested by the organizations developing such weaponry and tactics. The process is stringent enough that at the end, its author will be able to say, "I can assure you with a high degree of confidence that the risk of collateral damage is X, risk of compromising technology is Y, risk to U.S. systems is Z." [3]

The general nature of the chapter, which derives from a study of potential attacks on IADS, cannot hide its provenance. Yet, its lessons apply more generally. The concepts of strategy and defense are closely coupled, and militaries tend to be attacked not simply because they are easy marks but because they are targets of those who wish to diminish their capabilities any way they can.

Accordingly, we discuss the following challenges: getting in, doing something useful when there, evading counterstrategies, and assessing the damage afterward. Based, in part, on these results, we ask whether information warfare is ready for war.

4.1 Getting In

Consider three types of systems.

The first type can be accessed from the outside world by normal users who enjoy a full range of applications (for example, they can surf the Web). These systems may screen users by using passwords, digital signatures, or tokens. Users, for their part, are likely to have access to the outside world via e-mail, instant messaging (IM), or Web surfing. Such networks

---

[3] David Fulghum and Robert Wall, "Information Warfare Isn't What You Think," *Aviation Week and Space Technology*, 154, 9, February 26, 2001, p. 52–3. More recently Lieutenant-General Bruce Carlson, the chief of the U.S. Eighth Air Force, was quoted, "If you allow us not to use [explosives] but to use non-kinetic effects in these areas, we can predict the following behavior [the nondestructive incapacitation of computers] . . . with some of our tools we believe we can predict [effects] with a great degree of accuracy." The article added, "8th AF specialists can already accurately predict short-term effects. Operators can shut down a communications network or turn off a computer system or cause surveillance of a critical area to disappear for a few minutes." David Fulghum, "Battling for Minds," *Aviation Week and Space Technology*, 161, 9, September 6, 2004, p. 52.

typify most institutions. Some are easier to get into than others. Attackers may find ways to impersonate users. They may send messages directly to the system via ill-guarded ports. Alternatively, they can induce users to acquire strings of bytes via e-mail or the Web, and these, in turn, can do funny things to user systems and from there beyond. Of these systems, little further need be said; there is an extensive literature on the topic of hacking.

The second type of system is connected to the outside world but not in the same way that a general-purpose computer is. It is designed to react to a specific set of messages and no other. In turn, it can respond to such messages in only a preset number of ways. The last chapter noted the usefulness of such systems for secure applications. If such systems actually do as they are designed, which is sometimes harder to achieve than it looks, then getting into such systems is a matter of looking for weak spots in this message/response pattern. It takes a certain degree of discipline to design such systems. Giving them general-purpose direction is often difficult, so they are consequentially inflexible in the face of changing circumstances and requirements.

The third type of system is one not designed to be connected to the outside world. Such systems might characterize very sensitive industrial facilities (for example, chip design rooms), but are, or at least used to be, quite common within the military world. They probably still are common within less-advanced militaries.

Hacking into this last type of system may raise a sweat. Nevertheless, there are many ways to try.

First, an attacker could attempt to seduce, suborn, or scare an insider – whose usefulness, in turn, is related to (1) the privileges she enjoys, (2) the degree to which her actions can be executed without supervision or the possibility of correction, and (3) the risks she faces from her actions. This last suggests why it is easier to persuade an insider to spy on a system than to corrupt or disrupt it. Recruiting insiders, especially when one is unsure which insiders qualify, is risky in that it tends to require person-to-person contact, and when recruitment fails, authorities are often alerted, raising barriers to the next attempt.

Second, the attacker could look for remote maintenance ports, such as dial-in ports. They exist to let vendors and support contractors diagnose and, in some cases, fix systems without venturing on site. But how many critical systems, especially in third-world militaries, come with such dial-in ports? After all, (1) third world militaries rarely use third-party contractors, (2) such ports violate the systems' air-gapped status, and (3) they may make little sense for a mobile system that rarely connects to land-line phones. Nevertheless if foreign militaries follow the U.S. military's trend toward relying on contractors, and future equipment grows more complex and requires constant professional maintenance, the proliferation of maintenance ports cannot be completely ruled out – despite the security risks. However, when the penetration occurs, the only thing the hacker might learn after all their hard work is how close the system is to needing repair – not entirely useless information, but not particularly helpful either. Such information is, after all, the point of having a maintenance port in the first place.

Third, attackers associated with powerful intelligence agencies might persuade systems vendors to build into their systems backdoors that permit entry or can be programmed from afar to malfunction on signal. But how real is this threat? Defense equipment tends to be the world's least globalized sector precisely because of such trust issues. Rarely will U.S. equipment be part of our enemy's arsenals. Foreign manufacturers may be reluctant to induce deliberate faults into their systems and then reveal as much to the U.S. government. Getting caught doing so – especially in the face of so much suspicion – is a kiss of death for further exports, or even sales to one's own forces and friends.[4] Exploiting corrupted electronics, for its part, requires some way for attackers to receive signals, and, in the case of a full-fledged computer network attack, transmit them. The former requires an antenna, the latter a connected transmitter of some source. Either may add unwanted complexity to systems integration. And after all that, the device may be electronically inaccessible anyway.

Fourth, systems with no wired connections to the world may nevertheless be accessible through wireless devices (for example, infrared, Bluetooth, Wi-Fi, and Wi-Max) that are increasingly part of office equipment

---

[4] One Swiss supplier of cryptographic equipment was reported to be a CIA front and, as such, installed backdoors in its systems. As a small company, the supplier stood to lose little other legitimate business, and cryptography is an area where the benefits from corrupting a device are obvious. See Scott Shane and Rowman, *Baltimore Sun*, December 10, 1995, p. 1A. For the company's denial, see idem, *Baltimore Sun*, December 15, 1995, p. 23.

and consumer devices. If so, one could jack into a freestanding defense system by placing a pickup device sufficiently close it. At least as of several years ago, for every three wireless local area networks (LAN), one was running completely open, one was running with only the highly buggy and easy-to-break encryption[5] (Wired Equivalent Privacy, or WEP) that comes standard, and only one was running with a solid layer of protection on top. Things are better now; the 802.11b standard is yielding to the better-engineered 802.11g standard. Furthermore, getting to within a few hundred yards of such a defense system undetected is hardly easy. Even systems consigned to cluttered urban areas have to be found as such, and if one really knew where it was to within a few hundred meters, blasting it with the usual means may be far easier, unless collateral damage were an issue.

Fifth, one could run a wiretap into it. To succeed without getting caught, however, one must (1) find the wire, (2) get a tapping team in country and then out, (3) put the tap on without creating a telltale temporary disturbance while doing so, and (4) establish and maintain power to a communications link[6] from the tap to one's own assets – again without giving anything away.

Sixth, some military systems, such as an IADS,[7] communicate with one another using radar-to-radar links. Nevertheless, they are vulnerable

---

[5] See Craig Ellison, "Exploiting and Protecting 802.11b Wireless Networks," www. extremetech.com/print_article/0,3428,a%253D13880,00.asp, September 4, 2001.

[6] The historic tap and recover sequence obviates the need for a channel but cannot be used for real-time destruction or distraction.

[7] It should be added that opposing IADS is of particular interest to air forces, who, in turn, have tended to be the most enthusiastic proponents of information warfare. IADS have thus come up in several stories of how defense systems in the field have been hacked. The story that hackers disabled Iraqi IADS in 1991 was mostly a myth. Nonetheless, here is the story: the United States was able to hack Iraq's air defense computers by slipping several cooked electronic microchips into a French-made computer printer smuggled into Iraq during Desert Shield. U.S. News and World Report, *U.S. News and World Report's Triumph without Victory: The Unreported History of the Persian Gulf War,* Times Books (New York), 1992, pp. 224–5. The chips contained a virus that disabled computer systems by making it difficult to open a "window" on the computer screen without losing data. Because a peripheral device is rarely the site of a virus, it seemed like a good entry point for insertion. The Kosovo story was better authenticated but it is unclear whether the information warriors themselves know the results of their actions. David Fulghum, "Yugoslavia Successfully Attacked by Computers," *Aviation Week and Space Technology,* 151, 8, August 23, 1999, pp. 31, 34.

only if there is no functioning alternative communications infrastructure (such as fiber-optics) and operators do little to secure such communications from jamming and interception (by using spread-spectrum/code-division multiple access [CDMA] methods, for example) or fail to protect messages by encryption. In an era when hard encryption is easily purchased, the possibility that important communications are left unprotected may boggle the imagination. Yet, encryption is often avoided because, in a rickety communications system, it is one more thing that can go wrong. Furthermore, it may not be easy to add digital signal processors and encryption devices to transmit/receive modules built and installed by others as black boxes. Builders, if not local, may be reluctant to install patches in a war zone.

There are certainly enough options available so that a determined attacker has no reason to grow frustrated and quit too early. But not one of them is particularly reliable and all require, one way or another, that victims be haphazard about information security as well as overly enthusiastic about the joys of networking.[8]

4.2 Mucking Around

Information warfare can be used for spying, disruption such as denial of service, the corruption of data files or data streams, and sometimes even distraction. These all differ greatly in terms of how likely it is that they will create significant effects, how useful they are if they succeed, and how soon the victim may realize what is going on. Detection, needless to add, only starts the process of recovery; it does not guarantee that the source of the problem can be identified, much less reversed.

4.2.1 Spying

Eavesdropping on systems tends to be easier than disrupting or corrupting them even if there is this problem of bringing the information home from a disconnected system. Because eavesdropping is not meant to affect the

---

[8] Incidentally, the use of electronic warfare planes to capture what IADS operators see on their screens *is* possible without such intrusion because unprotected screens radiate energy (the Van Eck effect) – but one still must get close enough.

system's operations, it tends to be less detectable as well, and thus it can go on for a long time. Keeping a channel open requires that the code that opens the channel not be detected. Defenders, for their part, could authenticate every piece of software and vet all outgoing message addresses to guard against rogue code.

In practice, detailed authentication is hard to do on any modern general-purpose operating system; many, perhaps most, software anomalies are discovered by accident or because they produce side-effects. Taps may be eliminated by system rebuilds even if they are not discovered beforehand. Transmissions that go to strange addresses, go out at strange times, or are not squelched when everything else is (for example, during a test or when running in blackout mode) may give the tap away. The value of spying depends on what the target system is allowed to know. If it is a full-fledged node, then tapping into it may be like tapping into any other full-fledged node (for example, acquiring unit location, operational plans, weapons capabilities, and intelligence, surveillance, and reconnaissance [ISR] knowledge). If not, then what hackers discover may be no more than the machine at issue needs to know. For example, hackers may discover the machine's location, performance record, countermeasures (against stealth or electronic warfare, for instance), or performance parameters (which may, in turn, suggest its blind spots). They may also learn the radio frequencies that the machine uses or, better yet, how it hops among frequencies, or how it responds to higher-order information. However, hackers can uncover such information only if it sits in files rather than being hard-wired into the machine itself.

4.2.2 Denial of Service

Information warriors can stop systems in several ways. They can damage critical files upon which normal processes depend. They can create rogue processes (that is, an ongoing computer operation such as a data lookup) whose volume fills the system's available processor time, disk access capacity, or memory buffer space (as the February 2000 E-commerce attack and the May 2000 I-Love-You virus did). They may grab control over some essential facility in a system and thereby make it inaccessible. Most permanent of all, they might spawn a process that damages hardware either directly (by asking a component to execute a hazardous operation, for example) or indirectly (by establishing misleading conditions that cause humans to execute such an operation, for instance).

Some techniques, such as a flooding attack, work only as long as the malign process is active. If it comes from the outside, severing connections may alleviate the pain. Yet if the problem is recurrent (for example, the system is being recurrently attacked by a message-spewing zombie), detection and eradication may be more difficult. The site where the corrupt process sits may be unknown – and may not feel ill effects from the problem itself. If the problem is serious enough, the overall systems administrator may conclude that the entire network that has an ill component conduct some sort of reboot and file scrub – but this fix often requires that the network be completely disconnected from the rest of the world in the interim.

Denial-of-service attacks are a blunt instrument; they may disable the entire system or its components for a certain amount of time, but the more obvious they are, the more likely that operators would understand that the system has been attacked and react accordingly. Reactions range from attempting to understand and remove the problem to overriding systems manually that fail to perform reliably.

4.2.3 Corruption

By contrast, corruption may go undetected by its victim. Bogus data can look correct (for example, one fudge factor looks much like another). Complex system performance is, anyway, not expected to be 100 percent. A certain percentage of false alarms or false targets are to be expected. Something unusual may not be suspected until expected performance deviates substantially from actual performance and more common sources such as human error have been ruled out.

Yet corruption is not easy. A corrupted message must be well formed and well placed to be a message at all. By contrast, a disruption can arise from any single failure in any number of processes, depending on what parts of the system are how redundant. If an attack is to achieve its planned havoc, it must accord to the logic of the system's information flow (for example, command-and-control messages, log files). Many things have to go wrong in consistent ways. Otherwise, bad information in one file may be inconsistent with, say, five other files elsewhere whose existence was unknown to the corrupting hacker. For instance, altering a message that says that ten units were moved out of inventory last Tuesday may lead to inconsistencies with messages that list inventories based on a correct accounting of flows. It is often very hard to figure out which processes make use of which files and thus which downstream results based on older and more valid data are incompatible with newly altered upstream results. Original input streams (such as messages) can be corrupted with fewer such problems but only if there are no other input streams that reflect the original data.

Corruption also has to be plausible lest people question it, but a good deal depends on how much is, in fact, questioned. Sometimes inquiries are obviated by conditions of urgency, one-way links, and the inculcation of an "ours is not to question why" mentality – which easily translates into "ours is not to question whether." Corruption must also vie against bounds-checking algorithms similar to those that indicate that certain credit-card purchases are suspect. Conversely, while systems should be engineered against false anomaly, doing so may cause them to miss the true anomalies that signal enemy activities.

Defenses against corruption of existing files are relatively straightforward if the possibility of corruption has been contemplated in advance.[9] Someone using a device that combines a stand-alone ROM-based cryptographic element from a trusted source and with a keyboard or another form of manual entry can type and post cryptographically secure messages. Conversely, it is not always easy to assess the authenticity of messages generated by processes, even user-owned processes, rather than users. It requires being able to pass the authentication from a person or a device through software without being itself corrupted. It is probably simpler to design systems to take no important decisions except those based on directly authenticated messages.

---

[9] The device would need to be protected following the same tenets described in footnote 29 in Chapter 2.

4.2.4 Distraction

This fourth consequence of information warfare is rarely accorded the attention the other three have, largely because it tends to give the gloss of respectability to what appears to be the trivial effect that most skeptics accord to hacking: merely annoying the victims. Nevertheless, it is not inconceivable that hacking into systems can affect the performance of real-time systems (such as war machines or critical monitoring stations in civilian life) by distracting its operators, a process that counts on their emotional and, in some cases, physiological reaction to events not necessarily central to the functioning of defense systems.

Some distractions, such as creating false alarms, can be useful if they feed the tendency to ignore subsequent warnings (this trick has often been used in various crimes). As it is, lights sometimes flash when they should not. Some people fix the lights, but others live with the warning sign and ignore the risk that the next fault will be unable to announce itself before something more serious occurs. One would imagine that operators of defense and other critical systems would be more conscientious, but they may have far more warning lights to deal with than automobile drivers do. Warning systems may be easier to attack than the systems they warn about; they may be specifically designed to be accessible to those who monitor them. Yet, the gains from attacking warning systems may be thin and uncertain. Warning systems are often opaque to the outside. Any system design philosophy sophisticated enough to guard against "normal accidents" through an elaborate warning mechanism should have paid similar attention to information security issues. Too predictable a pattern of malfunctions may suggest that information attackers got inside the system; if the malfunctions are too well timed, they may tend to reveal exactly when the system needs to function to avert a threat (such as when strike aircraft are overhead or something else bad is about to happen). Finally, how is the attacker to know that a warning system is, in fact, being ignored?

Will all those warning bells, ringing phones, crank calls, blinking lights, unexplained animal noises, and curious lights in the sky actually distract operators? It takes little reflection to realize that warfighting not only requires attention, it compels it; the stress of war goes well beyond mere annoyance. A person may be distracted the first or even the second time. But sooner or later people adjust and so the value of such annoyance, while not zero, is still low. One might further imagine that hypersensitive and thus easily annoyed people are not generally selected to operate expensive machinery in the first place.

4.3 Countermeasures

Systems administrators in general and those warfighters among them in particular tend to be a conservative lot. They know things go wrong in peacetime, and more frequently in wartime. They hate single-point failure and often go to great lengths to find literal and metaphorical escape routes.

Although the primary defense for any system is attention to security, two overarching approaches – redundancy and learning – complement what security there is. Both provide a hedge against the occasional tendency of even the best information security to be tripped by undocumented features in system software.

4.3.1 Redundancy

Communications can profit from redundancy: redundancy in equipment and lines, redundancy across media (both e-mail and telephone, for example), redundancy within messages (such as encoding them so as to recover messages beset by bit error), and redundancy among messages themselves. Critical information can be reflected through multiple messages – for example, movement and location data are reported separately. Redundant conduits protect against denial-of-service attacks. Redundant files and processes protect against corruption attacks. Redundancy can provide some protection against exploitation (by permitting a message to be split and sent out over two unrelated media or at least lines, for example). But it also multiplies the number of flows or nodes that can be tapped.

Software redundancy is a trickier defense. To reduce the speed at which manual mischief (such as a hacker hopscotching among sites) or automatic mischief (such as viruses) spread throughout an integrated system, an administrator can limit connectivity and interoperability (via heterogeneous operating systems, for example), but the system pays heavily for such limitations.[10]

Hardware redundancy is nice but expensive. The goal is to design a system to ride out failure – especially induced failure of specific components of the sort that an attack in cyberspace might produce – without compromising performance parameters. One way is to favor systems with a large number of distributed capabilities over those with a small number of concentrated ones, if everything else is nearly equal.

The trickiest form of redundancy is of command and control – against the possibility that individual subsystems are either cut off from or, just as bad, have reason to doubt in the authenticity of messages they get. In a combat zone, where connections are falling down all over the place, alternate commanders could be designated to take over when the real commanders are cut off, compromised, or destroyed, *if they know this to be true.* It helps overall command and control if the alternate commanders are, themselves, linked into higher echelons.

Redundancy buys less if the problem is one of corruption rather than disruption and no one is sure whether anyone has tampered with the command messages. Indeed, actions by overly eager alternate commanders may only add to the confusion.

4.3.2 Learning

The maxim, "once burned, twice shy," applies in spades to information warfare, which, after all, is about deception, the susceptibility to which depends greatly on the victim's experience. Although the effects of information warfare can linger on in the tentativeness with which information is acted on, the risks of deception are mitigated by one's already having

---

[10] Matthew Williamson, a researcher at Hewlett Packard's Information Infrastructure Laboratory in Bristol, UK, has developed a method to retard the spread of viruses. The method flags processes that send out an unexpectedly high number of packets and limits the rate at which they connect to new computers while informing the authorities. See Matthew Williamson "Resilient Infrastructure for Network Security," www.hpl.hp.com/techreports/2002/HPL-2002-273.pdf.

been deceived. No information warfare campaign *that depends on deception for its effects* can be guaranteed to work forever. Like surprise, it is best saved for when it is most needed.

Knowing *that* one has been deceived hardly means knowing *how* one has been deceived, although they are related. One might know that a machine failed because one of its inputs was corrupted but not how such corruption took place. This partial knowledge nevertheless helps. The error, if induced again, may no longer lead to operational faults. Instead, there may be new procedures, such as checking the inputs manually or using three independent inputs, to keep such errors from creating hazards so easily.

How can one tell, for instance, that information has been corrupted? Sometimes, it is obvious, such as when a phony message is later repudiated and revealed as fake. Sometimes, it requires later investigation, and even after the problem is discovered, accidental or software-based flaws have to be distinguished from induced flaws. Nevertheless, the knowledge that some information is always subject to corruption, from whatever source, should lead users and designers to treat the information with more wariness. They could react by tightening security in general (scrutinizing password implementation or resetting default security levels, for example), by instituting redundancy (such as requiring separate processes for generating similar information) and bounds-checking algorithms, or by installing man-in-the-loop features (such as one that requires someone to check something before firing). As a general rule, knowing something could be wrong permits one to alleviate much, perhaps most, of the direct damage over the long run. As the last chapter argued, though, the very possibility of information warfare leads its potential victims to divert resources to defense, whether or not any attack is actually taking place.

Disruption is more obvious, but not how and why disruption occurred. Careful analysis and auditing may be required. If holes are found, they may be plugged; when systems fail, it is often because there are patches left to be installed in them. The mere assumption that a system has been disrupted may lead to policies that harden it, such as scrubbing the list of passwords and removing factory-installed defaults or anything that cannot be traced to an active user. At least such efforts may minimize the consequences of disruption (by duplicating critical data forward rather than relying on the reliability of communications channels, for example).

Information warriors, for their part, clearly have an interest in retarding such learning. Hackers routinely erase log files. Less routinely, they put misleading evidence into computers. In wartime, corrupted systems can be destroyed (for example, a missile tricked into going astray will rarely be recoverable for fault analysis). In a war, so can its victims.

Finally, militaries may learn from others. Iraqis helped Serbia defend against U.S. air strikes during the Kosovo campaign. Yet, international learning should not be assumed. Since information warfare is, at its heart, about deception, and most people do not like to admit to being deceived, they may be reluctant to talk to strangers about the experience. The potential recipients of such learning may figure that the capacity to be deceived characterizes others, not themselves. Nevertheless, given the growing prominence of information warfare in the U.S. arsenal, there will be great interest in any hard news on the topic.

To be sure, a successful attack may leave a useful residue even if the attack is detected and the vulnerability found and fixed. From that point forward, victims are less apt to believe what their machines are telling them. This cannot help but retard their reaction time when new events flash on the screen – but how much is hard to say.

In general, information warfare, as a weapon, has a strong tendency to depreciate once it is used, people are aware of its use, and potential victims adjust how they use and manage their systems in response. Short-run effectiveness may not say much about long-run viability.

4.4 Damage Assessment

Accurate battle damage assessment (BDA) on a target helps indicate whether attackers need give it further attention and what the prognosis is for operations (such as air raids) that require the system (such as an IADS) to be up. Knowing which efforts led to what effects permits attack resources to be allocated more effectively. Unfortunately, information about the effects of information warfare, besides being intrinsically hard to obtain, is itself subject to information warfare.

Iron bombs beget smoking holes and ruined structures that suggest how much damage the attack did. The less physical the attack, the less the certainty there is that it did harm. For electronic warfare, attackers have to be satisfied with simulated evidence (such as laboratory runs that indicated that the attack should have worked) and with indirect evidence (for example, attacking aircraft were painted [hit with radio signals] by surface-to-air missiles less frequently).

But both are easy compared to assessing battle damage in the wake of a hacker attack. Running it in the lab says nearly nothing because success is likely to depend on hard-to-predict flaws of the target system and how well it is administered. Indirect evidence is less specific than it is for electronic warfare.

Everything is made more complicated if the victims suspect they have been attacked and figure out what the attacker expected to see as evidence of success.

Prima facie evidence of successful *exploitation* would be a supply of information files and flows with useful information. Success ought to be apparent, especially if such information is corroborated elsewhere – for example, if the defense system yields its location and sensors are tasked to look for it there, they should find it. One defense against eavesdropping in cyberspace is to generate bogus information[11] that might mislead eavesdroppers unaware that what they have stolen was, in fact, planted for that purpose; one example was the World War II British ruse in which they planted a corpse carrying phony war plans to be discovered by the Germans ("The Man Who Wasn't There"). This bogus information can also be information about the system itself (such as phantom servers on the network) planted to force hackers to traverse an extraordinary profusion of paths to find the right door into the system. A more effective but less well-used ploy is to arrange for the hacker to learn false information

---

[11] These thoughts were triggered in part by the unpublished work of RAND colleagues Robert Anderson, Scott Gerwehr, Jamie Medby, Jeff Rothenberg, and Robert Weissler. Deception must be used with care. If it is overused, attackers are less likely to be fooled and they may look for indications such as unexpectedly low rates of file-updating activity in among the cabinets they are riffling through. Beyond some point, the intrusiveness of and the labor involved in creating deceptive material could easily deceive users more than foes.

in general (for example, phony deployment plans). At best, the ultimate recipients of doctored information will make poor decisions; even if they discover their error in time, they will be less apt to trust stolen information thereafter. Conversely, if the attacker has several taps into the system, the information going to each of the recipients should be consistent – so all the taps have to be found lest the eavesdropper grow suspicious upon finding inconsistencies.

The effects of *disruption* may be direct: certain nodes fail to communicate. It may be indirect – the system fails in hitherto rare ways, leaving the problem of correlating effects observed to faults caused. Because disruption is often obvious, the victim will know what happened and needs to nail down why it happened, so as to rule out accident or physical damage. This is the easy part; the hard part in deceiving the attacker is (1) knowing when one has been attacked again, (2) guessing what effects the attacker is looking for (which often requires knowing who the attacker is), (3) foiling the attack, and (4) generating evidence that makes it look as though the attack were carried out successfully. If the attacker can be convinced that it can use such deception to take down the attacked system at will, it may decide not to carry out more physically destructive attacks. Having been spared more reliable blows, the victim's system may then perform better than expected the next time, lying in wait until opposing forces have overreached themselves. All this presumes that (1) the victim has a capability that can substitute for the ostensibly disrupted capability, (2) there are important distinctions among physical raids that the victim can make in advance, and (3) the attacker has more confidence in its own information warfare prowess than is currently likely to be in evidence. All this considered, victims are better advised to bulwark their own systems and not get too cute about exaggerating the ostensible damage caused by information warfare.

As with disruption, evidence of successful *corruption* may read in consequent errors in adversary system performance even as other functions appear to be taking place normally. More directly, attackers can review the corrupted file to see that it is still corrupted – but this hardly guarantees that the victim simply did not discover the corruption and route around it while leaving the file in place. Can attackers be certain the altered file was the right one? Can they know whether the altered information was the basis for poor decisions or was ignored as implausible or inconsistent with other information? Here, too, victims can spoof the attackers, but doing so may not be worthwhile. Yes, people and processes that use the file can be subtly redirected to clean ones, but the greater the corruption (such as a change in a time-phased deployment schedule rippling into loading manifests), the greater the cost of redirecting everything. Deception of this complexity requires that the defender know exactly what information is used by what processes – and even then, attackers may have numerous undetected taps to catch inconsistencies that may indicate that the corruption has been detected. Perhaps those who understand their own system so well should not have been victimized in the first place.

Alas, it is too easy to react to insufficient BDA information by attacking repeatedly, but doubling the effort may reduce the effect. Although more attacks may mean more penetrations and alterations, it can also give away the modes, methods, and motives of the attacker. The target, once informed, can plug holes, work around them, or, as noted, send back misleading BDA. The odds of disclosure rise rapidly if making greater effort requires using progressively less expert hackers because the best ones are already fully employed.

4.5 Prediction

Intelligence on the target is the sine qua non of information warfare, but unlike more conventional forms of attack, it is essential in predicting even in broad terms how effective information warfare is likely to be.

4.5.1 Intelligence Is Necessary

Prior knowledge is more likely to suggest how far an attack can penetrate than how such a penetration can disable a system.

Systems attached to the Internet can be readily probed. Probing will often reveal which routes or ports are open and the kind of equipment and operating system being used. Deeper probes can reveal certain characteristic faults, such as unpatched vulnerabilities. Active probing can shed light on the ability of systems administrators to respond to intrusions. It may also unearth messages discussing the just-made probe and its ramifications on the attacked system. If probers are really lucky, they can also pick up the traffic on modifications made to the system in response to earlier probes. But such information may also be false, and reacting to it may clue defenders into a tap in the system.

Can one simulate the hardness of a system to information warfare by populating a laboratory with machines that the enemy is expected to own and with people of similar technological skills and command cultures (for example, trust among echelons)? Accurately portraying a red team was a staple of, say, air-to-air simulation. But by the time the Cold War started, Russian pilots had already been observed in two world wars. Monitoring their flying habits in the Cold War years added confidence to simulation designs. By contrast, opposing systems users and administrators are just getting into the network age. Probing an IADS system with decoys and feints was a common trick employed by the Israelis in the 1982 Bekaa Valley campaign and by both sides in the Cold War. The same probing carried out while the IADS has been tapped into should reveal even more about the effects of penetration on system performance.

The usual caveats about collecting intelligence on the security of information systems bear repetition:

- Savvy systems administrators will place a premium on hiding or disguising such knowledge. Probing may yield misinformation rather than information.
- Information warfare is Heisenberg country: once an attacker gets beyond routine port scans, the very act of collecting on an information system may alert its owners of inordinate hostile interest in them. Deep reconnaissance may trigger echoes that reveal holes to systems administrators, who can then plug them and thus change the vulnerability of the target. Conversely, peacetime probing may mask the discrete actions of potential foes preparing for wartime. Furthermore, sufficient activity may convince users that the system is untrustworthy, thereby persuading them to adopt less efficient information-handling techniques (even if the probing itself reveals nothing useful).
- Information systems operate in wartime differently than in peacetime. Since security often comes at the expense of usability, an optimal peacetime trade-off is for less security. In wartime, users, aware

that their discomfort is merited by circumstances, are less likely to revolt and more likely to acclimate themselves to their hassles.

Precisely how vulnerable a system is to penetration will depend on the details of its architecture, configuration, and management – details that not only differ from one owner to the next, but often from one month to the next. One would need to know how humans would react to failures in their machines – what trust would they put in them, what measures would they take in the fact of induced doubt, how redundant would machines be made, how quickly can they learn that something is wrong, and so on. There is almost no empirical data on acts of information warfare in wartime, and only scattered information on broader systemic relationships between computer-based information and the decisions they inform. This is not a form of warfare that can be parameterized based on physical principles (such as measuring radar cross-sections).

Predicting when an attack might work is important if the objective is to disable a system (such as an IADS) at a critical time (for example, when attacking jets are overhead). Hackers entering a system for the first time can only take a wild guess as to when they can get past the system's defenses – and hackers may not get a second chance if the effects of the first change are obvious. Hackers who work indirectly, by propagating viruses and worms, despite not knowing *whether* such an attack will disable the machines, at least have the comfort of knowing *when* such an attack might work. This comfort is of recent vintage. It took most of a day for the I-Love-You virus, for instance, to propagate worldwide. Four hours was required for the Anna Kournikova virus,[12] and the Code Red Worm

---

[12] Robert Lemos, "Fast-Spreading Code Is Weapon of Choice for Net Vandals," http://news.com.com/2009-1001-254061.html, March 15, 2001. Faster times have been reported in laboratory experiments. Dorothy Denning, "Cyber Security as an Emergent Infrastructure," in Robert Lathan, ed., *Bombs and Bandwidth: The Emerging Relationship between Information Technology and Security*, New York (Free Press), 2003, pp. 25–48. Denning's article cites Stuart Staniford, Gary Gim, and Roelof Jonkman, "Flash Worms: Thirty Seconds to Infect the Internet," *Silicon Defense,* August 16, 2001: "At U.C. Berkeley, a researcher showed how a 'Warhol Worm' could infect all vulnerable servers on the Internet in 15 minutes to an hour. Researchers at Silicon Defense took the concept further, showing how a 'Flash Worm' could do it in thirty seconds." The Witty worm, released in March 2004, infected more than twelve thousand hosts worldwide in 75 minutes. See www.cc.gatech.edu/~adjumar/witty.html.

had similar propagation times. The Slammer worm of January 2003,[13] however, had doubling and propagation times measured in minutes.[14]

4.5.2 Intelligence Alone Is Hardly Sufficient

Yet, not even the best intelligence on opposing systems can provide perfect predictability of the consequences of an attack in cyberspace. It is easier to forecast the outcome of a chess game because at least the board and the pieces do not change as the game is played.

Why the doubt?

First, IW takes place in cyberspace, a medium entirely capable of being changed by the defender. By contrast, physical or even electronic combat is constrained by all sorts of physical laws, such as gravity, atomic chemistry, or electromagnetic physics.

Second, the state of an operating system can change, on its own, from one microsecond to the next in ways that make it more or less vulnerable to particular attack patterns. Specific stack configurations that permit buffer overflow attacks can have very short life spans. The race conditions that may permit hacker-generated processes to read and write files normally off-limits to them are also measured in microseconds. Thus, a system's behavior under the same set of stimuli may not yield identical or even similar results.

Third, the same factors that frustrate the ability to verify, and thus secure, complex software systems make it hard to know how systems behave in the face of anomalous and potentially destructive inputs. A system is more likely to fail against attack not so much because of its overall architecture (although it plays a role) but because one or another faulty detail (such as a buffer overflow possibility) has been missed by the defender but found by the attacker.

---

[13] See Robert Lemos, "'Slammer' Attacks May Become Way of Life for Net," http://news.com.com/2009-1001-983540.html, February 6, 2003; Steven M Cherry, "Internet Slammed Again," *IEEE Spectrum*, February 2003, www.spectrum.ieee.org/WEBONLY/wonews/feb03.slammer.html.

[14] The Slammer worm resonated loudly through South Korea's broadband networks in the course of breaking out into the rest of the world, suggesting that these networks are reservoirs for worms and viruses in much the same way that South China is a reservoir for various forms of influenza viruses.

Fourth, the human element plays an enormous role in IW in terms of (1) detecting, responding to, and learning from attacks, and (2) being more or less resistant to deception.

Although it can be difficult to rewrite operating systems and critical applications over the course of a war, it is not impossible given competent programmers and access to the source code. The software for one U.S. air defense system – the Patriot missile – was, in fact, rewritten during Desert Shield to increase its ability to counter SCUD missiles.[15] A large percentage of all attacks that succeed in the real world do so because available software patches were not yet installed.

Changing the architecture of a system can also throw off a determined attacker. Networks can be rewired, routers might be added, internal addressing altered, and new internal barriers created to interfere with the attacker's ability to be certain that the right target was attacked. Simple backups could be used to guard against some information warfare attacks. Confirmatory messages could be passed via secondary channels (such as secure voice) and used to check other information.

Firewalls and other filters may be present. Copycats of the I-Love-You virus attack failed to cause much damage because systems administrators and Internet service providers (ISPs) learned to filter them out. Follow-ups to the Code Red worm encountered a far smaller population of unpatched machines running Microsoft's Internet Information Server software. Probing a system to test for a firewall too vigorously may create the very conditions that prompt systems administrators to install firewalls in the first place. Because the exigencies of operational security may mean far tighter controls in wartime than in peacetime, information gathered in peacetime may be inaccurate reflectors of what a wartime attack may accomplish.[16]

---

[15] There were other software errors, notably in the timing mechanism, that kept the Patriot missile from engaging the SCUD that struck the U.S. barracks in the late stages of the first Gulf War. See U.S. General Accounting Office, *Patriot Missile Software Problem*, February 4, 1992, GAO/IMTEC-92-96; B-247094. The software patch that could have fixed the timing algorithm was not delivered until the day after the attack.

[16] System security would likely be tightened as military conflict approached. The kinds of changes necessary to protect a system against "cheap attacks" would certainly be within the grasp of a motivated system administrator.

How, in fact, will systems administrators react to attack? At what point might they understand that their system is clogged? Do they have other devices they can use to control sensors and weapons if workstations are infected? Can workstations control sensors and weapons without being on the network – and after what delay (for example, from rebooting the workstation from fresh media)? Can such sensors and weapons be controlled manually, and if so, with what loss of effectiveness? How quickly can the network itself be purged and reestablished either locally within firewalls or globally? And since the systems administrator may well guess that the IW attack was deliberate (athough it could be just another virus), will steps be taken to make the attack appear to have succeeded even if its effects have been reversed?

Precision and predictability are simply unattainable in the face of such questions.

4.6 Is Information Warfare Ready for War?

Warfare is the management of violence, not merely its generation. With rare exceptions, it entails the organization and orchestration of a diverse set of means (that is, combined arms) to achieve definable and often measurable effects. Mobs and terrorists can be counted on to ignore these rules in seeking the unrestrained expression of passion or the creation of spectacle almost as ends in themselves. But they are rarely a match against a professional military, should it come down to a confrontation.

Professionals in warfare understand that before a device or technique can be considered weaponized, multiple hurdles needs to be surmounted:

- Command and control
- Predictable effects and collateral damage
- Conformance with recognizable norms of conduct
- Deployability in time and space
- Integration into combined arms
- Safety in storage and use
- Integrated logistics support
- Training

As an instructive example of why this may apply to weapons, even the most powerful ones, consider atomic bombs. Trinity proved that the physics was valid. Prior mass air attacks on Japan's cities gave them a still-controversial basis in warfighting norms. But to turn nuclear devices into nuclear weapons required, at a minimum, that the device be designed and built so that one could be put on an aircraft that could fly from U.S.-controlled bases to the target. For instance, the original atomic bomb was built to fit the internal dimensions of the B-29 bomber. Many of the World War II campaigns in the Marianas were motivated by the need to secure air bases sufficiently close to Japan. Even then, atomic weapons were used primarily for their immense psychological effect and only secondarily to reduce Japan's military capability. After World War II, the Department of War turned to other important features of their management: creating a nuclear establishment to increase warfighters' understanding of their effects, promoting safety in their storage and transportation, building a logistics infrastructure to handle them, developing doctrine for integrating them into overall operations, and training people in their use and management. Issues of command and control pervaded the entire nuclear era.[17] The same issues will matter for hostile conquest in cyberspace as well.

4.6.1 The Paradox of Control

The applicability of all warfare – and thus offensive information warfare – necessarily depends on a certain level of control over one's own efforts and, if successful, over the options faced by opponents.

Paradoxically, because signal can be manipulated with much better fidelity than can noise, information warfare may be antithetical to the ethos of war. True, at the tactical level the ability to impede the enemy's use of its own resources has value regardless of how it is achieved. But at any level, both the precise effects and the control of warfighters can be challenging. Reducing the control that an adversary possesses renders

---

[17] See, for instance, Paul Bracken, *The Command and Control of Nuclear Forces*, New Haven (Yale University Press), 1983; Bruce Blair, *Strategic Command and Control: Redefining the Nuclear Threat*, Washington (Brookings Institution), 1985.

war-limitation, as opposed to warfighting, strategies *harder*, not easier, to carry out.[18]

Exercising operational command and control of information attacks may require more care than it first appears. Military personnel, if they are assigned to be the attackers, are disciplined individuals, but a large part of what makes people adhere to discipline over time and in crisis is the sense of being watched. To exercise such control in information warfare requires being able to monitor a cadre of attackers – even down to the keystroke level – and to prevent information about a target's network architecture or vulnerabilities from leaking to those outside the circle. Loosen oversight – something implied in the increasing popularity of mission-type orders – and pretense of exercising any control whatsoever vanishes. The very difficulty of tracing attacks back to their origin, coupled with the amount of detailed information held by attackers and not passed upward, frustrates normal command and control. Judgments are constantly being made often in real time because doors jarred open in cyberspace may be closed, literally within seconds. Thus, discretion has to be available to attackers.[19]

4.6.2 Other Weaponization Criteria

Next on the checklist of what it takes to make a weapon is *conformance with recognizable norms of conduct*. Computer network defense is certainly on safe grounds, even if active defenses – counterattacks automatically

---

[18] Thus, consider the wisdom of attacking nuclear C^2 systems, especially those systems that are rudimentary (such as those of India, Pakistan, or China) or newly fragile (such as Russia's) to begin with. Countries unsure of their C^2 reliability (and, except for Russia, facing the far larger U.S. arsenal) may react quickly and unpredictably if they fear losing control of their nuclear forces.

[19] The ability to go after fleeting targets argues for giving combatant commanders broad operational authority to conduct whatever information warfare operations are necessary. Nevertheless, until the place of information warfare is understood to be one among many weapons, the rationale to have all targets cleared by the White House has validity. Nuclear weapons were originally considered large weapons and only later acquired their strategic role (as deterrence rather than warfighting systems). Information warfare will not mature until it makes the opposite transition – downplaying the strategic possibilities to focus on their tactical utility. That may be happening. William Arkin has listed nearly a dozen offensive information warfare programs. William Arkin, *Code Names: Deciphering U.S. Military Plans, Programs and Operations in the 9/11 World,* Hanover (Steerforth, 2005).

delivered back to the attacker – are not entirely acceptable.[20] Computer network exploitation is hardly more offensive than the kind of eavesdropping that routinely occurs with wireless communications. But computer network attack is by no means established, especially when directed against civilian infrastructures (such as the bank account of Slobodan Milosevic). Several years ago, Russian Foreign Minister Igor Ivanov tried to convince United Nations Secretary-General Kofi Annan that the destructive potential of computer network attack was tantamount to that of strategic nuclear weapons, and therefore it should be banned by treaty. Many Russians writers conclude that Moscow would respond to a serious IW attack much as it would to a nuclear attack – with a nuclear counterattack.[21] Not for nothing has the release authority for information warfare, as with nuclear weapons, been vested in the national command authority.

The same dilemma holds for *predictable effects and collateral damage.* Again, as discussed, knowing what an attack did is hard enough; predicting as much is even tougher. Perhaps the risk of collateral damage to *other* systems from an intrusion into a *specific* system is exaggerated. However, an important and often overlooked source of collateral damage is that the affected system may be controlling or influencing processes

---

[20] According to Brett Stalbaum, the creator of FloodNet, "[Pentagon programmers] were redirecting any requests coming by way of the EDT Tactical FloodNet to a page containing an Applet called, 'HostileApplet.' This Applet . . . instantly put all the FloodNet protestors' browsers into an infinite loop by opening a small window which tried to reload a document as fast as [possible]. . . . I had to restart my [computer] to recover control." Dorothy Denning, "Activism, Hacktivism, and Cyberterrorism," in John Arquilla and David Ronfeldt, *Networks and Netwars,* Santa Monica (RAND, 2002), pp. 239–88.

Although it is hard to feel much sympathy for the attacker, an active defense has three practical drawbacks:

- It legitimizes an activity that it is not in the interest of the world's most wired societies to legitimize.
- It carries the risk of great damage to third-party systems that the hacker may have gone through in order to conduct the attack (and may give the attacker motive to work through a third party, such as a newspaper, to put the defender in the position of offending the third party and suffering the public consequences).
- Even if an attack succeeds, the assets that an attacker has put at risk may be easy to replace, at worst (for instance, computers can easily be purchased for a few hundred dollars these days).

[21] Matthew Campbell, "'Logic Bomb' Arms Race Panics Russians," *Sunday Times,* November 29, 1998, as taken from Stephen Blank, "Can Information Warfare Be Deterred?" *Defense Analysis,* 17, 2, 2001, pp. 121–38.

and services of which the attacker is unaware. Attacking, say, the Global Positioning System (GPS) not only degrades the performance of U.S. weapons systems but adds hazard to civilian transportation.

With their effects so difficult to predict, *deployability in time and space* also becomes a practical barrier. In theory, computer hackers can operate day and night with effects carried around the globe. In practice, the susceptibility of an information system to attack depends critically on its details. These details not only have to be discovered to determine whether a target system can be penetrated and with what effect, but they may vary without notice and from one microsecond to the next. What works today may fail tomorrow, and visa versa. Not only is considerable effort, often spanning months and years, required to scope adversary systems, but the efforts must be constantly renewed in order to validate their relevance at the time of attack. And even then, the descent into crisis and war is often the moment that they change the most, as users suddenly begin to take their security much more seriously.

Such uncertainty makes *integration into combined arms* a hairy bear of a problem. If one cannot predict the effects of information warfare on the adversary, one cannot begin to trade it off for or synchronize it with other forms of warfare. What would, for instance, an air planner change about his strike or his attempts to suppress enemy air defenses (SEAD) if computer hackers were confident that their efforts silenced the threat? Probably very little at this point. Furthermore, at least within DoD, some of the problems with integration are self-inflicted. The practice of lumping all the various facets of information warfare together suggests that these facets will be integrated with each other before they are integrated with more kinetic forms of warfare. This is not necessarily the best sequence of integration. An information warfare attack designed to confound an adversary's logistics system may yield better results if coordinated with air strikes on that system than with a psychological warfare campaign.

At least *safety in storage and use*[22] is relatively straightforward, unless the rogue code is stored in the adversary's system. Ironically, certain aspects of computer network defense bear more watching, if for no other

---

[22] Safety in use refers to the risk to the operators, and not from the risk of friendly fire – which is not a trivial concern for information warriors.

reason than some of them are constantly running. Among these are active defenses, intrusion detection systems, viral countermeasures,[23] and deception techniques of the sort that risk confusing users as well as attackers. The risk is not necessarily in their design, but in the possibility of error in their implementation that lends them undesirable system behaviors.

Integrated logistics support for information warfare is not problematic as such but cannot be ignored. The problem is not so much movement – information warfare operators, tools, and techniques can be deployed anywhere – as maintenance. Hackers must stay abreast of the state of the art in information systems, a process that requires constant validation, verification, and assessment.

Finally, *training* is an area that receives requisite attention within the armed services and is also not per se problematic. Indeed, information warfare is blessed by the fact that so many civilians are adept at it and that readily observed performance gaps separate the professionals from the merely passable.

All in all, information warfare in the form of computer network attack has a long way to go before it is fully fit for the front.

4.7 Conclusions

Militaries generally pursue technological developments because they see their adversaries becoming more formidable. Precision munitions development was emphasized to counter the Soviet Union's seemingly unstoppable mass of armor. Radar was a reaction to efficient attack airplanes. Stealth development followed the success of Soviet-supplied batteries against the formerly invulnerable Israeli Air Force. Iraq's use of pop-up SCUDs hastened attention to theater missile defense and systems to get inside the enemy's five-minute engagement cycle.

---

[23] Aficionados of Linux had boasted that the system has only known two viruses: the first, which caused trouble, and the second, propagated to eradicate the first. However, add-ins to the Linux kernel are not without flaws. In mid-2002, the Linux Slapper worm infected several thousand servers by creating a peer-to-peer attack network; it exploited a flaw in the mod_ssl module for the Apache Web server. See Robert Lemos, "Linux Worm Causes Peer Pressure" www.news.com 2100-1001-958122, September 16, 2002.

Information warfare is exceptional enough to warrant skepticism about its ultimate efficacy. Effort is spurred by the possibility (or hope) that adversaries planning to use information technology to improve their capabilities will so bungle the job as to permit information warriors to get into previously inaccessible systems and disable them from the inside.

Desert Storm, for better or worse, demonstrated that sometimes adversaries *are* less clever than we feared. One can only contemplate with glee the possibility that future defense systems will be built atop an operating system[24] with built-in facilities for denial of service: "Sorry, you changed your hardware parameters and the system is shutting down. Please call to reregister your machine." Alternatively, such systems might be engineered to allow friendly "technical support" people to take over your device entirely. It could happen.

Yet, it is best not to plan on such things. If for no other reason, everyone is becoming more conscious of the information warfare threat to critical systems. The truly paranoid may even come to rely on systems that hardwire everything they need to work. So a plausible lower limit for the amount of damage that information warfare can inflict on its targets is zero.

---

[24] Early plans for Microsoft Windows XP would have had computers running the operating system to tell Microsoft some of their attributes (e.g., memory size, attached devices) as part of the registration. Such data would then have been embedded in the operating system. If the software were subsequently run on a machine with different parameters it supposedly would conclude that it had been copied to another machine (in violation of its license) and would then shut itself off. This led to fears that any substantial change in a computer running XP would trigger the software to disable itself. See "Inside Windows Product Activation," A Fully Licensed [GmbH] Paper, www.ilicenturion.com, July 2001. The license agreements of many of Microsoft's update packages (Service Pack 3 for Windows 2000 and Service Pack 1 for Windows XP, for instance) have the user acknowledge and agree that Microsoft may automatically check the operating system's version and automatically download updates to the machines. Brian Livingstone, "Sneaky Service Pacts" *Infoworld*, 26 August 2002, p. 20.

5

Information Warfare against Command and Control

When tales of command and control gone bad are told, the moral of the story is often how hard waging war is without full information.[1] Commanders ask themselves: where are my forces; what are they doing; what is their condition; have my commands been correctly received and interpreted? Those commanded have their own concerns: do they have the correct orders; do such orders reflect the situation on the ground; and how do these orders compare with what their cohorts have been ordered to do?

Such questions grow more urgent when communications nodes are under attack. Units out of touch cannot receive orders or report back to command. Militaries consequently lose their coherence, and therefore their efficacy. Much of the credit for the relatively easy victory of Desert Storm was ascribed to Iraq's inability to exercise command and control after their communications links were attacked.[2] Information warfare advocates argue that computer network attack should have a similar effect on digital data networks. It is all about information scarcity.

But what if information scarcity is becoming history? What if the real challenge of command and control is how best to cope with a glut of information? What would information warfare look like if it reflected

---

[1] See, for instance, Martin Van Creveld, *Command in War*, Cambridge (Harvard University Press), 1985.

[2] See, for instance, Alan Campen, *The First Information War*, Arlington (AFCEA Press), 1993; R. W. Reading, "Could Iraq Have Made Better Use of Its Air Force and Missile Technology during the Air War?" *Australian Defence Force Journal*, 94, May/June 1992, pp. 39–63.

as much? To limn a possible answer, this chapter therefore (1) explains why the era of scarce information is ending; (2) the problems this may cause and how people may cope; and (3) how to think about strategies to explore, exploit, and exacerbate the dysfunctional aspects of information excess.

5.1 The Sources of Information Overload

One need not study the use of technology very hard to understand that information scarcity hardly describes the modern economy or even modern combat, or to understand why this is the case.

Not only are communications and geospatial information becoming better and cheaper, but also more robust thanks to the profusion of diverse communications media. A world once limited to land-line phones and walkie-talkies has burgeoned into one with Internet broadband connections through cable, digital satellite links (DSL), very small aperture terminals (VSAT), multipoint microwave distribution systems (MMDS), laser-based point-to-point links, fiber-optics to the home,[3] and maybe even high-bandwidth electric power cords. Cell phones come in several flavors besides the standard second-generation digital models that used to be typical in North America: Europe's GSM, PCS (Sprint's service, for example), second-and-a-half generation (introduced here by Voicestream), third-generation CDMA service, direct-connect systems (such as Nextel), cordless phones that boast a range of nearly a mile, and space-based phones (such as Iridium, Globalstar, and Australia's Optus). Hybrid systems include infrared wireless LANs, Wi-Fi (802.11b), Wi-Max, and Bluetooth-enabled devices. Militaries use satellite, microwave, and high-frequency systems and are experimenting with unmanned aerial vehicles (UAV) as communications carriers. Communications can be and increasingly are protected from jamming by digitization coupled with sophisticated algorithms for propagating signals such as frequency-hopping, CDMA waveforms, and ultra-wideband signaling – all protected by other

---

[3] Verizon introduced its fiber-optic service (FIOS) in 2004. See Ben Charny and Jim Hu, "Verizon's Fiber Race Is On," http://news.com.com/2100-1034_3-5275171.html, July 19, 2004, 12:13 P.M. PDT.

error-correction algorithms (such as trellis encoding) that are, for that reason, attack-resistant as well. Each may be defeated – but can all of them be, simultaneously?

Geolocation methods – critical to answering the question, "where are my forces?" – are also proliferating. There is GPS, of course, plus fair prospects that Russia's Glonass satellite navigation system will recover and evolve, and that Europe's Galileo satellite navigation system will join them circa 2012. Local positioning systems may work (via pseudolites, for example) if global systems are unavailable. If one has a reliable initial vector of one's location, continuous updates are available from increasingly better inertial navigation systems (through the use of fiber-optic and ring-laser gyroscopes, for instance). Finally, more of the world is becoming quite accurately mapped in digital form, permitting location to be estimated by a combination of mapping and dead reckoning; toting the world in maps 50 gigabytes at a time is no problem.

True, this hardly proves that communications will never henceforth constrain operations. War planners would be foolish to assume that the bandwidth required for this or that weapons system (for example, UAVs) shall flow forever free from effective assault. Requirements also have their way of filling the bandwidth available – or beyond – if imagination exceeds the limits of Mother Nature. A battlefield strategy based on videoteleconferencing everywhere has to come up with reliable bandwidth from somewhere, and wishing it so will not make it so. The commander's playthings always vie with a soldier's needs.[4] Single-point failures can affect systems with ill-considered physical architectures – for example, routing telecommunications via common structures such as microwave towers or common channels such as railroad rights of way. They can also affect systems with feckless syntactic architectures (such as poorly backed-up directories). A systematic investigation of a communication grid's true dependencies and bottlenecks by an information warrior is always required before passing judgment on it. Nevertheless, militaries

---

[4] Even in so recent a war as Operation Iraqi Freedom, soldiers with very little connectivity to the Global Information Grid were still detecting the enemy the old-fashioned way – by running into them. See David Talbot, "How Tech Failed in Iraq," *Technology Review*, November 2004, pp. 36–45; it is based on unpublished RAND work by John Gordon et al.

with the sense to put command-and-control information (such as compressed voice, text, and perhaps some symbols and graphics) atop the requirements pile and do a first-order systems engineering scrub should find that the bits get through one way or the other.

5.1.1 Its Effect on Conventional Information Warfare Techniques

The efflorescence of information content and conduits tends to complicate the three basic goals of information warfare: denial of service, interception (exploitation), and corruption of information files and flows.

The more pathways exist, for instance, the more difficult disconnection and denial of service become. True, denser networking allows certain types of mischief (such as worms) to propagate faster, magnifying the damage before countermeasures can be brought into play. Nevertheless, the physical and technological diversity of communications paths complicates a strategy of targeting a single point – a key consideration when the subject is command and control.

Interception, one would imagine, would, in fact, be easier when there is an information glut. After all, with more conduits, any one fact is more likely to be accessible by more people. Thus, more pathways are available to be searched, multiplying the odds that one can be tapped and made to leak. But just as one side can drown in information, so can those who would intercept such a flood; indeed, this is a growing problem within the intelligence community. To see why, divide information into three categories:

- Critical information that needs to be known only by few (such as a spy's identity)
- Critical information that needs to be known by many (for example, war plans)
- Noncritical information (pizza orders, for instance)

Information in the first category can, should be, and usually is compartmented and carefully guarded. Furthermore, there is no inherent reason that improving information *technologies* should lead to a rise in the *amount* of critical information in existence (for example, the names of every secret agent). Really critical information should never see a computer; if it sees a computer, it should not be the one that is networked; and if the computer is networked, it should be air-gapped.

Guarding the second category is made more difficult by a profusion of information technologies, but it is not impossible. Careful decomposition of information in terms of who needs to know what can ameliorate the risk.

A hard rock mining analogy may help explain the third category. When mining costs decline, it is the production of low-grade ore[5] that rises. High-grade ore would be mined in any case. Similarly, the information generated and disseminated when information is cheap is that which previously would have been deemed too worthless to collect. As more of *that* information is intercepted, the percentage of all information in the top two critical categories goes down. So long as most costs were for mining and few were for smelting and refining (to continue the metaphor), the economics of interception are only slightly bent. But, the U.S. intelligence community, of late, has found that the costs of extracting and assessing its take are rising faster than the costs of collection. If the cost of doing analysis – defined as finding the needle in the haystack – rises in proportion to the size of the haystack, then the efficiency of interception must perforce drop as information piles upon information.

The same holds for deception, but for somewhat different reasons. One might think that the difficulty of altering a key perception of the adversary would hardly be affected by the presence of so much noncritical information. But noncritical information (for example, what soldiers tell one another) is what gives critical information (such as the army's morale) its credibility. Deception works to the extent that corroborating and validating information is either similarly altered or made unavailable. The ability to fool the Germans about the Normandy invasion was greatly helped by Germany's inability to pull much information out of England.[6]

---

[5] To be precise in the use of metaphors, oil and gas extraction works differently. Almost everything discovered is worth pulling up. The response to higher prices is to dig holes in progressively smaller, more remote, and more speculative reservoirs.

[6] See Michael Howard, *Strategic Deception in the Second World War*, New York (Crown), 1990; John Cecil Masterman, *The Double-Cross System in the War of 1939 to 1945*, New Haven (Yale University Press), 1972.

The more data are around, the greater the gap between falsely induced beliefs and the mass of evidence that supports them.

Despite these generalizations, the efficacy of any one attempt to disrupt, intercept from, or inject false information into systems varies greatly from one case to the next. With the possible exception of interception, whose harvest is a deep, dark secret, no one really knows how well information warfare would work in the real world against serious adversaries. But there are good grounds for believing that information glut makes life harder for conquerors of cyberspace.

5.2 Coping Strategies

If information warfare would confound an opponent's command and control, then it must take today's information overload into account. One must first ask what a superabundance of information does to command and control *absent* information warfare – and the first place to ask is what it does to the logic of allocating authority, because such logic should reflect who needs what information.

5.2.1 Who Makes Decisions in a Hierarchy?

How should information overload change where decisions are made? To start, ask why decisions are made at the command levels that they are.[7]

---

[7] This question presumes that decision-making authority is rationalized in terms of decision-making efficiency, but historically there have been many other reasons that decision-making authority is distributed as it is. Many combat units, for instance, were owned by their commanders (for example, kings, barons, the wealthy), who could retain decision-making authority even for minor matters. Even when militaries became national institutions, commanders were often from a higher social class than the commanded and arrogated for themselves the privileges that went with status. Legitimacy can be even stronger when commanders are selected on meritocratic principles; even if everyone knows the same facts, commanders earned their authority by demonstrating themselves to be wiser and more capable. Necessity and sometimes success also conferred commanders with an aura of authority or charisma – two very important qualities if one is to make men go in harm's way. Conversely, there are reasons to move command down into the ranks: (1) people perform better when allowed to exercise initiative, (2) creating the opportunity for poor decision making accelerates the culling of bad officers and bad ideas, and (3) decapitation has less effect when lower levels are used to making their own decisions.

When it comes to who should have the right to make decisions, commanders retain two key information-based advantages over those they command. First, it is usually safer to reserve highly privileged information to a few commanders than to a mass of lower-level officers; more ears mean more leaks. Decisions predicated on such information cannot be made except by the few who know. Second, to the extent that good decision making requires the coordination of various components, top-down decisions are easier to make. Compared to a committee of peers, a commander is more likely to be informed by a single set of goals and beliefs whereas peers may be competing among one another for greater glory or less risk. Command is also a more straightforward method of resolving contention than negotiations.

Are there countervailing information-based factors[8] that call for pushing command down the hierarchy? Historically, difficulties in communication explained the need for local initiative. Lower-level officers invariably knew much that upper level commanders did not,[9] but what they knew could not be communicated whenever there was too much new information developing too fast to stick into any available communications pipe – were one even available in the first place. Even when the pipes were available, there remained the difficulty of making meaningful use of such information. A hundred units reporting location or fuel status is not particularly actionable information without some way of getting it to tell a story. Maps and spreadsheets help, but more commonly headquarters had to work with markers, acetate, stubby pencils, and accounting sheets, all slow and hard to manipulate. With technology, today's pipes and tools are better.

Information technology, therefore, should shift authority from the field to headquarters. Such a belief encouraged the rise of conglomerates of the 1960s (such as ITT and Gulf and Western – corporations that

---

[8] Fans of complexity theory claim that optimization algorithms based on centralized data processing can be outperformed by processes that model the various actors as independent agents, who are given a set of rules and programmed to respond to each other. Even were it so, nothing prevents the commander from using a distributed-agent model rather than a conventional linear programming model as a decision-support tool. One should not confuse technical mechanics with institutional arrangements.

[9] For instance, in the U.S. Civil War battle of Antietam, Union soldiers won the battle only by ignoring the commanders and taking initiative at Burnside Bridge.

managed heterogeneous divisions solely by looking at the financials). In a military context, reliable communications could acquire a stream of information on unit location, status, morale, and contacts. Video-teleconferencing lets commanders look subordinates in the eye and feel they have a sense what the front looks and feels like. Accurate geospatial information helps one build large, instantly refreshed pictures of how units were positioned and where they were moving relative to each other, especially if aided by the automatic extraction and mapping of location data from messages. The image of the commander surrounded by video screens and manipulating the various chess pieces into play certainly appeals to commanders themselves. It is also entirely consistent with the art of the possible – if they can determine which feeds matter.

The ability to amass and convey more hard data from the field, however, is not the same as receiving perfect information. Many of the softer bits of information cannot be evaluated without some sense of what lower-level officers mean when they say something. Some are whiners; others are deniers. Some are more articulate than others. Some have more credibility, based perhaps on a track record that reflects luck more than anything else. Intuition is not always easy to convey. Morale is particularly difficult to assess without being there. To some extent, however, the noise that is added by those who convey information has to be compared to the noise that enters everyone's decision-making process because of biases and preconceptions.

So, why shouldn't decision-making authority float upward? To some extent, and much to the discomfort of warfighters, it has. Examples include the White House direction of the Mayaguez rescue, the vaunted ability of Air Force generals to talk directly to fighter pilots over Kosovo, or the micromanagement by Central Command out of Florida that characterized the early portions of the Afghanistan campaign. Yet, none of these examples are brilliant testimonies to the virtues of top-down command. The quality of perception that comes from being on scene does not always translate well over the wires. But then, this is not a problem that can be dealt with through more reliable communications and greater bandwidth. Because information systems can do little to improve matters, information warfare is unlikely to do much to worsen them.

The greater problem with centralization is one of limited attention. Attention is often called the only persistent scarcity in the information age.[10] Commanders can, if they choose, now flood themselves with information in ways they never could have before. It is not that the quality of information is bad. Even perfect information would pose difficulties. The real problem lies in how well humans, because they are humans, can react to complexity. There is only so much time in a day. The number of linked concepts that a person can remember and transmit has been famously represented as "seven plus or minus two."[11] It is hard to work more than one problem at one time. Commanders foolish enough to try to micromanage the battle themselves risk being overwhelmed in short order. Key decisions would not be made on time, or, if made, would not gain the benefit of due consideration. Nothing so frustrates a new perspective than the inability to focus long enough on whether one's mental model corresponds to a changed reality – or whether, instead, the model needs to be fixed. Over the long run, few things are as crucial in war as the ability to learn.[12]

To some extent the existence of more knowledge is driven by the growth in what needs to be known; life and especially war keep growing more complex. There are more weapons systems, each of them more complicated than the last. The possession of information systems not only compels attention to what they contain but also to how these systems are to be fed, managed, and protected. Everyone's ability to collect terabytes of what might be labeled intelligence is much greater than it ever was. The ratio of support costs to warfighting costs rose steadily throughout the twentieth century. There are more occupational categories because there

---

[10] See, for instance, Michael H. Goldhaber, "The Attention Economy and the Net," http://www.heise.de/tp/r4/artikel/6/6097/1.html, November 27, 1997; Josef Falkinger, "Limited Attention as the Scarce Resource in an Information-Rich Economy," www.iza.org/index_html?lang=en&mainframe=http%3A//www.iza.org/en/webcontent/teaching/bonn_research_seminar/bonnseminar_description_html%3Fsem_id%3D876 & topSelect=teaching&subSelect=bonnresearchseminar, November 10, 2004.

[11] George Miller, "The Magical Number Seven, Plus or Minus Two: Some Limits on Our Capacity for Processing Information," *Psychological Review*, 63, 1956, pp. 81–97.

[12] See, for instance, Richard Overy, *Why the Allies Won*, New York (W. W. Norton), 1996; Eliot Cohen and John Gooch, *Military Misfortunes: The Anatomy of Failure in War*, New York (Free Press), 1990.

are more specialists – that is, people expert in things that commanders are not expert in. Since modern warfighters cannot be treated like cannon fodder, proper attention must be paid to their health, both physical and psychological, as well as to their finances and families. When it is impossible to keep all this stuff in one's head, the role of the specialist rises apace; work is delegated to staffs.

There is no going back. So how do we cope?

5.2.2 Responses to Information Overload

Most of us intuitively know our own limits in dealing with too much information, even if not all of us have the discipline and the confidence to appreciate and deal with that fact.[13] The strategies that people use to cope with the flood of information – including, of course, delegation – have therefore gained importance as the water levels have risen.

We can start with what will probably not be chosen: reverting to yesterday's thin and shaky communications systems. Ignorance has few advantages in a wartime context. Even if it did, people are rarely rewarded for withholding information on the grounds that it might cause clutter. Instead, they are often penalized for holding back information that just might have proven useful. Nor can commanders necessarily duplicate the filtering techniques that others, with limited channels, used to decide what was worth sending up the chain of command. Knowing one's own command needs is not the same as conveying useful judgments about what lower levels should find important. Besides, sometimes the importance of a fact is obvious only to those in the field who know it.

---

[13] Geoffrey Parker argues, for instance, that Philip II was a micromanager – not always the best way to run a multicontinental empire. *The Grand Strategy of Philip II*, New Haven (Yale University Press), 1998. See also Eliot Cohen, *Supreme Command*, New York (Free Press), 2002. As has been more recently reported:

> To deal with the problem [that the agent is not a double agent deliberately feeding misinformation], the CIA has adapted a 'no threshold' policy in which it takes all reports seriously. But this, [former Directorate of Operations officer Thomas] Carroll says, 'only *exacerbates* the disinformation problem,' and has already prompted the House Permanent Select Committee on intelligence to call for modifying the 'no threshold' approach (Vern Loeb, "Beyond the Blame Game," *Washington Post*, September 9, 2002).

Having lower-level operations, logistics, and intelligence officers anno-tate their messages and indicate what *they* feel is important is not only work on their part, but adds its *own* clutter to what they send forward. Expect them to cheat, anyway. People tend to compete for attention, and so inflate the importance of what they have to say. Many label everything as urgent, and then more urgent, and then most urgent to draw the requi-site attention to one's issues, much as advertising has drained significance from words of praise. In the old days, the very ability to send a message at all spoke to its urgency and importance. Spam is computerdom's way of telling us this is true no longer. The very ease of communications has shifted the burden from senders to recipients.

So what methods, shortcuts[14] as it were, *do* people employ to cope with information glut?

Delegation and the hierarchical decomposition of work is the most straightforward way. In the past, subordinates had freedom of action because no one could *gather* the information required to know enough to tell them what to do. In the future, they will have some freedom of action because no one had the time to *process* the information required to tell them what to do. There are major differences between the two, encapsulated in Martin Van Creveld's term, "directed telescope" – the commander's discretion to focus on one sector of the battle. While this technique was generally a sound reallocation of resources as the fight in this or that sector became more or less critical to the final outcome, it never overcame limits on how much more information commanders could actually receive from the sector; telescopes can only do so much. Today, the information is all there to begin with, but the problem of context switching is, if anything, worse. A sector may become active literally within seconds now, compared to hours in the past. Worse, everything touches everything else, if for no other reason than weapons ranges are longer, so that several different types of units may engage any one target. The ability to direct a telescope suggests why micromanagement is more prevalent in small-scale contingencies and peace operations than it is

---

[14] "As the stimuli saturating our lives continue to grow more intricate and variable, we will have to depend increasingly on our shortcuts to handle them all." Cialdini, op. cit., p. 7.

in high-intensity warfare. Perhaps one reason that Desert Storm was conspicuously not micromanaged from the White House was that, with everything taking place simultaneously, it simply could not be.

Another coping strategy is to rely on distillations of information from a specialist as opposed to a subordinate. Some can be summoned to advise on a specific problem and are thus stafflike. Others advise on general issues, such as what tomorrow's weather will be like, how the mind of the enemy works, which products have what kind of performance, and what the advantages and disadvantages of certain operational techniques are. Dense communications networks permit them to serve everyone at once. This alone does not let the commander off the hook. Experts often evaluate decisions based on their expertise without respect to broader considerations. Delegating problems in advance to selected experts works only if broad categories of concern are unrelated to each other. Thus, personnel, logistics, intelligence, force protection, and so on are dealt with one at a time, without great need to contemplate the effect of one, such as logistics, on another, such as personnel.

Clutter can be reduced through managing by exception, concentrating on a handful of parameters, and attending to the rest only when performance becomes unsatisfactory – a dollop of optimization for a few variables mixed with a barrel of satisficing[15] for the rest. Doing this well, though, requires not only the wisdom to separate the chestnuts from the gravel, but also some degree of luck. One must hope that warnings will show up in the less well-watched parameters while there is still enough time to act – without sending up too many sparks in such quick succession that one hardly knows which to attend to first. Perhaps needless to add, if the unattended parameters are represented by specific people (for example, by slighting logistics, one slights logisticians), their tendency to emit distracting warnings to gain attention must also be considered.

Alternatively, one could ignore unlikely contingencies or, better yet, not dream them up in the first place – this despite the fact that in wartime,

---

[15] "Satisficing" is a word invented by Herbert Simon, a Nobel Prize–winning economist, to describe the behavior of individuals who make choices not on the basis of what is best, but what is good enough.

clever enemies do odd things precisely because they are unexpected. Such contingencies are otherwise discovered by identifying the assumptions used in one's analytic construct and subjecting them to continual reexamination[16] – so perspective is acquired by not engaging in such a process. In that way, one can avoid having outcomes shake fundamental beliefs, among which have historically been that victory goes to the offense, that spirit rather than technology decides wars, that people freed from their dictators will be grateful and compliant, or that yardage gained is a good proxy for progress in wartime. Perhaps needless to add, the reverse of these tenets have also been assumed even when unwarranted. If victory is elusive, then "no matter – tomorrow we will run faster, stretch out our arms farther . . . and one fine morning. . . ."[17]

The last simplifying assumption is that the world is static, or, more specifically, that one's enemy is static. After all, if strategic objectives and military culture tend to remain fixed over long periods, how much of a stretch is it to assume a foe's tactical goals and operational tendencies to be similarly invariant? Such an assumption is a comfort. Hard as it is to assess one's actions by first gauging the enemy's reactions, it is far more difficult to contemplate that the enemy's reactions will take yours into account. Unfortunately, next to failure to learn, the second most fundamental mistake in warfare is failure to anticipate that the other side also learns.

Such coping strategies are not always the simplifications of fools. By and large, many decisions *are* best off being made by others, not every issue merits equal attention, second-order effects are lower-order concerns for a reason, and fundamental assumptions do not change from day to day – not in reality, and not in the minds of one's opponents. But every effort in the direction of cognitive simplification comes at some cost and risk. When the contest is close or where small advantages cumulate into large ones (as in the OODA [Observe, Orient, Decide, and Act] loop), then such costs and risks may be telling.

---

[16] See James A. Dewar, *Assumption-Based Planning: A Tool for Reducing Avoidable Surprises*, Cambridge (Cambridge University Press), 2002.

[17] F. Scott Fitzgerald, *The Great Gatsby*, New York (Scribner's), 1925, p. 182.

Yet, these are only the *rational* responses to information overload. Many responses, while preserving emotional stability, clearly lead to worse decisions.

Some people, for instance, cope with information overload by concentrating on what they are familiar with; after all, they do have more information on the topic now. This comes at the expense of other topics, about which they are now showered with more information than they can know what to do with.

Others really cannot see the forest for the trees. The more trees, the more obscure the forest. People fixate on the specifics, factors that make every piece of information unique; they spend their mental energy trying to master details. This leaves them little scope or energy to rise above such details and look at what applies to all of them or discern patterns that become visible only as they relax their intense focus.

At a far end of the coping spectrum is simply shutting down; having absorbed all the information that one can, everything else – or at least everything else that does not conform to preconceived constructs – is simply ignored. Or, having taken information in but being plagued with doubt, people postpone decisions, fearing, paradoxically, that not enough or at least not enough reliable information has been received. Sometimes this is called "paralysis by analysis," but it predates the rise of formal analysis as such.

Figure 5 shows these responses as two groups. Some strategies make integration more difficult – that is, they favor the processing of information in separate containers and they do not encourage the insights that come from integrating knowledge. Other strategies minimize the possibility of change and therefore discourage agility.

Can more information technology cure the ills brought about by information technology? Intelligent agents are touted as having the potential to grow into good butlers – letting in the welcome guests (such as useful e-mails) and excluding others (spam, for example). As with many forms of artificial intelligence, we are "almost" there and will be for a while to come. Visualization technologies may help people array and thus absorb large quantities of certain types of information (for example, a large percentage of all messages refer to unit movement; these are more easily

Rational Responses	Irrational Responses	
Delegate to experts	Flee to comfort zone	*Inhibits integration*
Delegate down	Focus on trees, not	
Ignore 2nd-order effects	forest	
Manage by exception	Shut down inputs	*Inhibits ability to anticipate surprises*
Ignore the unlikely	Postpone decisions	
Assume static parameters		

Figure 5. Responses to Information Overload

displayed than read). Maps remain the dominant visual metaphor for war, and they have unique capabilities for storing and displaying large quantities of relevant information in real time (for example, DoD's Common Operational Picture). But if we rely on a metaphor, we risk being captured by it. Maps work well only for conflicts in which who sits on what terrain says something useful and primary; during the Vietnam War, maps were used as a crutch in the absence of a better understanding of the conflict.

5.3 Know the Enemy's Information Architecture

Whether attackers are better off trying to deny information to the enemy or feeding them more than they can handle depends on what information architecture the enemy has. There are systems (such as IADS) whose performance can be reduced through the successful disconnection of limbs from stem. But is this always the best approach to information warfare?

The following two subsections mull an alternative strategy, one that does not deny information to the enemy's command-and-control system but shoves information into it so as to exacerbate whatever dysfunctional mechanisms they have for coping with overload. Such information is pumped into the adversary's command-and-control system through cyberspace – but it need not be. Hard copy, gossip, the media, or irrelevant facts on the ground may have the same effect.

Information operators, must, however, first understand the enemy's command-and-control system in terms that would give them some feel for what the dysfunctional coping strategies might be.

5.3.1 Elements of Information Culture

One clue to how an organization will react to excess information is its information culture, the norms that govern how information is organized and shared. Here are some questions that may shine a light on this.

How freely does data flow? Are people more likely to hoard and then trade information or share information based on trust?[18] Does more information flow vertically or horizontally? Is vertical integration accorded more, less, or the same weight as horizontal collaboration as a way of putting the big picture together? Will people seek out knowledge even at the risk that it contradicts their earlier judgments – especially ones for which they already have argued?

Since data have to be given due credit before people make decisions on them, one must ask, what confers credibility on information? Are people more likely to believe what they get from public sources, or are they more likely to believe what their friends tell them? Does the spoken or written word convey more authority? Must decisions be justified on the basis of facts and arguments, or will authority, experience, or charisma do the job? To what extent are formal credentials taken seriously as a source of authority or credibility?

Knowing this will not necessarily be enough to predict how people cope with information overload. It helps to understand, if possible, the adversary's broader culture. Certain cultures may be more prone to

---

[18] Frank Fukuyama has argued that cultures can be characterized as high- or low-trust; see Frank Fukuyama, *Trust: The Social Virtues and the Creation of Prosperity*, New York (Free Press), 1995.

compartmentalize and then analyze, others to rest on simplifying assumptions about the overall gestalt. The coping process is also often highly individual; because it varies from person to person, there is no substitute for knowing the individual opposing commanders.

Answers will suggest, in broad terms, how information flows and thus what flows of information are apt to produce overload.

5.3.2 Elements of Nodal Architecture

The next thing to understand is the real organizational chart – not who reports to whom but who, or alternatively, what node, influences[19] which decisions? Nodes are typically persons, but they can be groups, committees, or similar bodies.

Influence can have many sources. Some comes from delegation, whether vertical, as is the case in hierarchies, or horizontal, as is more typical in more complex and networked organizations. Personal connections can confer influence on individuals or groups. Complexity fosters specialization, and specialists can have great, if narrowly circumscribed, influence even absent a title. Fixers and go-betweens can acquire unexpected sway. If key people can be identified and targeted, they may, themselves, be influenced and thus influence others.

Influential nodes may also be characterized from a process rather than simply a content perspective. If a node's influence comes from its being a potential bottleneck, information operators may see it as a place to induce congestion. Congestion, in turn, retards decision making. Defenders who ward off such effects by adhering to fixed decision timetables may preserve the speed of decision making but at the expense of its quality.

5.3.3 Injecting Information into Adversary Decision Making

Conversely, if there already is, in fact, too much information upon which to base decisions – or at least too much information of marginal value or dubious reliability – then bottlenecks can be forced by injecting

---

[19] See, for instance, Jena McGregor, "The Office Chart that Really Counts," *Business Week*, February 27, 2006, pp. 48–9.

information into critical nodes, rather than starving them of information. Just as humans have optimal weights, so there is an optimal weighting of inputs for decision making. Beyond that point, nodes must put more time into sifting and sorting – much as before, some point nodes must put more time into gathering. Both create bottlenecks.[20] Coping by arbitrary rationing of information does not restore the more tolerable erstwhile conditions if it knocks out good information to make room for the injected information of little value. Worse, when decisions have to be justified through recourse of facts, there may have to be some reconciliation between the facts available to the commanders and those available to the commanded – which in turn, may also be injected facts.

So, how can information warfare affect command and control?

One way is to inject information into the cyberspace of adversaries as part of a broader influence campaign. Knowing the enemy's commanders permits operators to tailor the message to the individual, which should increase its effectiveness. Where influence is wielded through cyberspace, information operators might hijack or create communications that purport to come from others. When information is excessive and attention is scarce, the resources to check the authenticity of influential nodes may be inadequate.

But there are limits. Tailored facts cannot vary too much from facts available to everyone else. U.S. political or advertising campaigns, despite decades of increasing sophistication, are still inexact arts – and that is with the advantage of copious demographic information on customers (as Chapter 8 discusses) of the sort rarely available on third world adversaries.

Hijacking a message – inserting a false one where a real one would be expected – is a potentially powerful variant of simple insertion. Many e-mail systems have little protection against message spoofing; witness the I-Love-You virus of May 2000 or the more recent advent of what are known as phishing messages (messages that appear to have been sent from financial institutions or electronic marketplaces but which were created to persuade users to divulge identification or account numbers and even

---

[20] By analogy, among the most hard-to-counter forms of computer network attack are those created by information overflow resulting from the actions of zombies in a distributed denial-of-service attack.

passwords). Yet, it is possible to create a messaging system that is nearly impervious to simple deception through the use of well-managed digital signatures. Because authentication systems do not always scale simply, in order to be truly secure, they must be kept small. A compact system would be necessary, perhaps limited to the command structure plus other influential nodes.

Secondhand information (for example, "he said this, I wrote") is not so well protected from mischief of all sorts, not just from tampering in cyberspace. Again, the impact of rumors will depend on people, not programs. Having received a bogus message from someone, will recipients exercise their suspicion to question it? If they do, how effectively can they discern the truth? To what extent can organizations be driven by rumor? Who trusts what from whom?

Generally, people can convince others that messages sent out under their authority are bogus ("no, that's not me – this is") as long as they themselves are aware of the possibility and use one or a variety of techniques – for example, maintaining a clean channel back, using cryptographic methods such as digital signatures, or issuing disclaimers with sufficiently personal, and thus authenticating, content or style. The proliferation of channels observed in the beginning of the chapter can be actively employed to purify communications as well. Nevertheless, a bogus message *could* be accepted without much thought. It could have verisimilitude and thus not have its verity questioned. Or, it may be so urgent that there is no time to contact the originator for a check. In either case, corrupted messages have to obey a key tenet of deception: plausible consistency.

Mischief may also be introduced by people outside the trusted hierarchy creating bogus messages. The pains taken to authenticate a source should be directly proportional to the credibility of the source in the first place; for example, leaders are authenticated and thus credible, and foot soldiers are authenticated but weakly and thus trusted but not entirely, whereas random voices are not authenticated and must be taken with a grain of salt. There are also sources that are poorly protected but trusted by people who know the person whom they purport to be. One of the virtues of cyberspace for potential deceivers is that communications along this medium are disembodied and are therefore more easily substitutable.

Finally, spoofers may hope for the residual benefit of impressions vaguely left behind even after the original source is shown to be a fake. So, operating against any information system but the most fully secure provides at least some scope for inserting misleading information.

5.4 Ping, Echo, Flood, and Sag

If command and control is heir to certain faults, especially from information overload, information operators may want to pursue a strategy of exacerbation. One component of this strategy, ping and echo, injects information into adversary information systems in order to explore its potential command-and-control dysfunctions. The other, flood and sag, uses the knowledge gained to exacerbate such dysfunctions.

5.4.1 Ping and Echo

How would the enemy react to information injected into its cyberspace? Attackers, injectors as it were, would be tickled most pinkly if their victims took these false events seriously, acted in reaction to them (such as by redeploying forces to the wrong location), and thereby created for themselves an unexpected military disadvantage. Historically, deception has worked just that way. There is no reason to believe that deception has lost its power, since it relies more on the victim's predisposition to believe something than on the technical power to simulate that something convincingly.

Such hopes aside, it may be worth pinging the system with bad information simply to hear its echoes in the victim's e-mail, instant messages, blogs, and documents.[21] Even the harmless scurrying that would take place as plans are invoked or authorities called would yield useful hints about the enemy's command-and-control infrastructure, its predisposition to action, concepts of operations, and perceptions. A ping that raises the blood pressure of a system may suggest how close adversaries are to information overload – and what shortcuts or simplifying assumptions they take to cope with it. For instance, does the commander, if stressed,

---

[21] John Hockenberry, "The Blogs of War," *Wired*, August 2005, pp. 118–23 and ff.

cease responding to certain messages of a certain parameter (for example, petroleum, oil, and lubricants [POL] supplies) or sector? Are possible contingencies brushed aside? Is unwarranted emphasis placed on details at the expense of broader strategic aims? The prospect of finding that the foe cannot cope with information overload in a very useful manner may justify a campaign whose feints catch the other side's attention, resulting in the foe paying less attention to other indicators and therefore raising the odds that they will go awry before they are noticed.

Admittedly, there has always been a lot of false information – "fog" – in every military system. Its presence per se connotes nothing extraordinary. But were victims to perceive false information as a test, and were they to have some idea which channels will then be monitored, might they be motivated to fake their reactions? Maybe not. Creating a consistent suite of false reactions, especially in response to unexpected deception, is difficult; even tamping down or exaggerating true reactions across an organization is not easy.

5.4.2 Flood and Sag

So, for the information warrior, addition may make more sense than subtraction. Indeed, addition is subtraction; adding random information reduces the information content of a data stream. This is particularly so if the targets already have more information than they know what to do with. Resources are already stretched trying to figure out what has value; raising the question of what is valid and worth notice adds further strain. If coping with information glut introduces distortions, adding to the pile should introduce further distortions (unless polluting an information source leads others to discard it and therefore alleviates, in a perverse way, the opponent's information glut). Incidentally, neither the source nor the conduit itself need be in cyberspace. Word of mouth, print, or mass media can all contribute to the pile. As such, computer network operations constitute one slice – albeit a growing one – of a broader IW campaign whose function is to confound enemy command-and- control systems or decision making in general. But ears in cyberspace – instant and unobtrusive – may be the sine qua non for monitoring the process and using the results to tune such a campaign.

Putatively, how an organization responds to injected information should parallel how it responds to an information glut. Such responses, in turn, suggest potential weaknesses that military operations can seek to exploit. This logic rests on several assumptions, all plausible even if none is guaranteed.

The first assumption is that responses to information overload can be characterized consistently across an organization. Because it is people who respond to stress, uniformity of response may not be absolute. But these responses should also be expected to reflect (1) institutional incentives and (2) organizational cultures – which tend to be more consistent. Culture, in particular, has a tendency to reinforce itself. People are trained in it and train others; believers in it are promoted while the skeptics quit or at least quiet themselves.

The second assumption is that if organizations react in a characteristic fashion to information glut, they will intensify their reactions if the glut grows. Thus, if organizations already tend to focus on exceptions at the expense of forecasting, seeking opportunities, or pursuing everyday optimization, then flooding it with information will make them tend to hold even more firmly to the proposition. It will focus perhaps on the exceptional exceptions and downplay every other facet of management.

The third assumption is that one can introduce sufficient additional information into an organization to make a palpable difference in its behavior – another way of saying the additional noise will not get washed out in the ambient noise.

So calculating, one might devise an information warfare plan that would:

- Collect general intelligence on the opposing organization to characterize its decision-making culture
- Plant surveillance mechanisms (notably in cyberspace) to monitor how people react
- Insert irrelevant information into the system
- Measure how the organization responds

Figure 5, which listed ten types of responses to information overload, characterized them as those that tended to inhibit integration and those that inhibited agility. At the risk of oversimplification, warriors in cyberspace may want to induce reactions that exacerbate such mechanisms thusly:

- If the victims react to information overload in ways that limit their ability to integrate operations, they may be more vulnerable to actions that would normally be met by the orchestration of forces or effects. This creates an opportunity for militaries that are good at parallel warfare or massively combined arms of the sort that would force enemies to respond with requisite complexity if they are to respond well at all.
- If victims react to information overload in ways that limit their agility, they may be more vulnerable to actions that call for adaptability. The victim's rigidity would reward carrying out operations that can be opposed only by forces with sufficient ability to learn. This creates the opportunity for militaries that are good at carrying out the unexpected or can exploit technologies that directly upset what its enemies perceived to be its capabilities.

5.5 Conclusions

Theories of command and control that focus on the problem of getting enough information to make decisions fit a world in which information is scarce. But information is no longer scarce. Instead, it is ubiquitous. Information overload may come to characterize organizations. Knowing how organizations cope is critical to understanding the effect of information warfare on them.

Were information scarce, an information warfare strategy to deprive opponents of information would be on the agenda – but not so if information and information channels are characterized by abundance.

A better strategy in that case may be to work with rather than against information trends by feeding low-grade information to the adversary with the hope of exacerbating whatever dysfunctional strategies it has adopted to cope with the influx of information. If that works, subsequent operations can be shaped to take advantage of them.

6

Friendly Conquest in Cyberspace

The observation that "All's fair in love and war," points to the double-edged meaning of conquest, as well. Conquest in both love and war is about the subversion of autonomy and triumph over personal will.

In war, there is very little ambiguity about the nature of the contest. Pace Clausewitz, the aim of conflict is to disarm the enemy. The tasks of conflict are often expressed as control over some medium: occupying terrain, sweeping ships from the open ocean, or claiming and maintaining air superiority.

In love, there is considerable ambiguity, self-deception, and thus no small amount of humor. But the results are often quite similar – a surrender of autonomy and the creation of dependence, which can be more or less symmetric among lovers.

Analogies apply to cyberspace.

Hostile conquest in cyberspace is about conflict. Typically, there are two sides. Defenders have an information infrastructure whose correct operations are matters of importance to, say, military security or making money. Warriors attack it from the outside (through flooding attacks, for example), but more insidiously from the inside (such as through intrusion). They aim to steal good information, implant bad information and commands, or just confound the system. Rarely is such conflict symmetric in the sense of like versus like; think, instead, of jet versus a surface-to-air missile (SAM). Permanent control over someone else's system is very difficult, and not always the point anyhow. Hostile conquest is more often expressed not as control but as denial – the demonstrated ability to take from the victim full use of its own infrastructure.

Friendly conquest in cyberspace is based on different mechanisms. It builds on voluntary transactions, at least at first. Yet, conquest can be said to have occurred if subsequent interactions and dependencies enable the conqueror to make reliable and effective use of the assets of the conquered. It can lead to unwarranted influence of the conqueror's systems – unwarranted in the sense that the victims' need for continued access to the conqueror's system exceeds what its underlying value to the victim would suggest.

This chapter characterizes friendly conquest in cyberspace by:

- Defining it
- Noting some advantages of coalitions as the conduit of soft power
- Examining how an enterprise's information infrastructure can shape coalition formation asymmetrically
- Providing some examples of asymmetric coalitions involving relationships with individuals
- Providing similar examples involving organizations

In fairness, the various examples are either constructs or else abstracted from examples of information exchange that predate the dense, intelligent, and interactive cyberspace discussed in this book. Friendly conquest in cyberspace, so far, is more evident in the world of packaged software than in enterprise-to-enterprise relationships, where owners remain wary of extending their cyberspace to create coalitions, much less subjecting themselves to the largely undefined risks of dependence. So, this and the next chapter rely less on argument from evidence and more on argument from logic as a way to display the potential of and the mechanisms for conquering cyberspace in a friendly way.

6.1 A Redefinition of Conquest

Like love itself, soft power in cyberspace comes in two-sided dance or three-or-more-sided contests.

In a two-sided dance, both partners believe that the relationship is beneficial and each seeks to wrest as much advantage from it as possible. One or both partners may be wary of dependence; each may fear that dependence would be unbalanced and thus a source of power by one over the other. But both enter the dance anyway.

A three-sided struggle is the competition of two to seduce a third. Its aim is less to best the opponent absolutely, as in war, but to better appeal to third parties. Politicians compete with each other for votes among the third party, the body politic. Various products seek consumers and operate in the knowledge that purchasing a product from one vendor fills needs that would otherwise require a purchase from another. Even opposing military combatants, locked in a two-sided hostile struggle with each other, often appeal for support from world opinion, the third party. The stakes are less obvious in cyberspace but no less complex.

In a triangle, one aim is to make it easier for you than your opponent to gain a third party's assets. A related but clearly distinct aim is to make the assets of the third party more usable by you than by the third party. It helps if the third party shapes its assets to conform to your requirements – for example, by collecting data that you are interested in and in the form that you can read and manipulate more easily. Interoperability at all levels is the goal. The close working relationship that the U.S. intelligence community has with its counterparts in Britain, Canada, and Australia may not be unrelated to the fact that they are the world's four largest English-speaking countries, language being the first level of interoperability. An added advantage with this approach is that the more of their information assets are oriented to how they do things, the more it will cost them to leave the coalition or at least that aspect of the relationship – and the more that others must offer to bid them away.

The logic of soft power in coalitions (or relationships) hence rests on the following tenets:

- Coalitions are important to success in a wide variety of endeavors
- Coalitions increasingly get value from the exchange of information
- Information exchange is frequently mediated through cyberspace
- Coalitions linked through cyberspace are often characterized by increasing, often unbalanced interdependence

6.2 The Mechanisms of Coalitions

As Thomas Friedman has observed,

We have moved from a world where everyone wants to go it alone – where the rugged individualist is the executive role model and the vertically integrated company that does it all is the corporate model – to a world where you can't survive unless you have lots of allies, where the Churchillian alliance maker is the executive role model and the horizontally allied company is the corporate model.[1]

The business literature concurs.[2] The oft-lauded Silicon Valley model refers to the drive of companies to dominate well-defined horizontal layers within their industry rather than offer a complete soup-to-nuts solution for its customers.[3] Domination requires alliances, or at least

---

[1] Thomas Friedman, *The Lexus and the Olive Tree,* New York (Farrar, Straus, Giroux), 1999, p. 184. Adds John Sculley (former CEO of both Pepsi and Apple), "When we talk about virtual corporations today, we're mainly talking about alliances and outsourcing agreements. Ten or 20 years from now you'll see an explosion of entrepreneurial industries and companies that will essentially form the real virtual corporations. Tens of thousands of virtual organizations may come out of this." From overseas, Bruno Lamborghini, Olivetti's economics director, stated, "A company's competitive situation no longer depends on itself alone but on the quality of the alliances it is able to form," "Corporate Odd Couples," *Business Week,* July 21, 1986, p. 100. And from a more academic perspective, James R. Lincholn, Christina L. Ahmadjian, and Eliot Mason remark,

> A consensus has emerged on the importance of inter-firm networks as a modern mode of organizing economic activity. Across an array of industrial settings, well-formed networks are enabling small and specialized firms to outcompete large, diversified, and integrated corporations. In today's global economy, where technologies and markets are volatile and uncertain, such partnerships offer a variety of advantages in speed, flexibility, efficiency, and even welfare. One such benefit that is attracting broad attention from scholars and practitioners is the capacity of networks to diffuse knowledge and information (James R. Lincholn et al., "Organization Learning and Purchase-Supply Relations in Japan," *California Management Review,* Spring 1998, p. 241).

[2] See, for instance, Ken Auletta, "American Keiretsu: The Next Corporate Order," *New Yorker* (1997), pp. 225–7. The vision behind CMGI, the web-holding company, was to create a network of interlocking Internet companies that worked together with each of CMGI's sites to feed its users into the others. See *Business Week,* October 24, 1999, p. 141.

[3] For a nice portrayal of the Silicon Valley way of business, see Charles Ferguson and Charles Norris, *Computer Wars,* New York (Random House), 1993, pp. 174–82. See also David Manasian, "Reboot System and Start Again, *Economist* February 27, 1993, pp. 7–11.

compatibility with other enterprises operating at other layers to succeed. Enterprises compete to dominate their niche, in part, by forming both tacit and overt coalitions with others.[4] Thus interfirm competition is often a matter of competing networks or ecosystems.[5]

Coalition formation is also a key aspect of *internal* management.[6] This was not always so. In hierarchical organizations, the existence and legitimacy of a chain of command permitted tasks to be successively decomposed into subtasks and ultimately assigned to individuals. With greater worker autonomy and the assignment of the more difficult problems to crossdepartment and crossdisciplinary work groups (task forces, tiger teams, networks, and so on), the ability to get nonroutine work done requires willing, even enthusiastic, internal partnerships. Such relationships work best if enterprise goals are strongly aligned with each member's individual goals and thus effort. Coalitions often organize the process of contribution; they assume greater importance as outsourcing blurs distinctions between who is inside and outside an enterprise's walls (as per Chapter 3).

The importance of coalitions applies even, perhaps especially, to war. The U.S. military can apply force anywhere with help from no one. But its ability to prevail in recent wars has been due, in no small part, to its ability to make common and effective cause with other fighters, such as the Kosovo Liberation Army or Afghanistan's Northern Alliance. By contrast, the paucity of help from others, not least of which has been hard-to-establish Iraqi security forces, has hurt stabilization operations in Iraq.

[4] John Harbison (of Booz-Allen Hamilton) reckons that some thirty-two thousand alliances have been formed around the world in the past three years, three-quarters of them across borders. "The Science of Alliance," *Economist*, April 4, 1998, p. 69. The article adds, "In Silicon Valley, alliances break up almost as quickly as they form, rendered obsolete by some new technological twist, or undermined by a small firm's worries about being 'sucked dry' of its good ideas by a larger partner." Even as early as 1993, *Digital Media* had counted 348 alliances in pursuit of multimedia services, a large share founded along the New York–Hollywood–Silicon Valley axis. "Media Mania," *Business Week*, June 12, 1993, p. 110.

[5] Damin Darlin, "The IPOD Ecosystem," *New York Times*, February 3, 2006, p. C1.

[6] Charles Handy, the British author of *The Age of Unreason*, even suggests that some corporations might become more like voluntary associations, run for the benefit for their working "members." "The Creative Economy," *Business Week*, August 28, 2000, p. 79.

True, military coalitions have yet to make much footprint in cyberspace. But with the spread of cell phones and their continuing transition to digital form, this footprint can only grow, and perhaps quickly. If and when it does, the ability of the U.S. military to attract its partners into cyberspace – the better to deal with them efficiently and to bind them to its way of doing things – may be of no small value.

The ability of militaries to prevail in future conflicts may thus depend less on what they own and more on what they can find, bargain for, patch together, and exploit. Increasingly this "what" is information, and increasingly cyberspace is the source and the medium for such acquisition. As described further in this chapter, urban conflict, such as that which has attended stabilization efforts in Iraq, depend on the ability to acquire information, and that ability depends in large part on what it can learn from the locals. Although it is ridiculous to suggest that DoD can harvest such information *solely* by setting up attractive chat rooms, instant messaging forums, and game spaces, such a proposition grows less outlandish as the population of Web-ready cell phones rises worldwide.

6.2.1 The Particular Benefits of Coalitions

People prefer to own resources rather than depend on others, however reliable and friendly, to supply them – but people do not always have such a choice. The United States, for instance, cannot exploit Europe's military potential by owning Europe, or even its militaries. Even where choices do exist, borrowing rather than owning other peoples' assets has a few things in its favor. One need not pay for their upkeep and management. Conversely, owning them may create opposition from potential rivals or from third parties such as governments. Success at wooing partners and getting to use their assets also signals one's desirability or at least inevitability as a partner.

Coalitions, in theory, put the power of all behind the interests of each. This is more obvious when partners share the same interests such as the defeat of a foe or the passage of favorable legislation. But it also holds when, say, a pool of assets is to be divided. Synergy and harmony among members help coalitions beat their counterparts. Conversely, divergent goals, ambiguities in command and control, unrationalized duplication, and the lack of interoperability often lead to coalitions being less than the sum of their parts.

Coalitions need not spread their benefits evenly. Indeed, the lure of disproportionate benefits is what often energizes core members to form coalitions. They may cherry-pick the capabilities of their partners to emboss or round out their own (for example, in a coalition of fleets, each navy may fight separately, but the efforts of each one's minesweepers may be usefully combined). In politics, achieving asymmetric advantage is often a question of whose interests are served first or are postponed. In economics, it is revealed by the distribution of gains. Some coalitions unite clear winners with others, who, had they exercised better foresight, would not have entered the coalition, or at least not under its reigning terms, but are now locked in. But such cause for regret is neither necessary for the winner nor, in the long run, particularly desirable. It is enough if the winner were markedly better off.

Finally, even if the synergy that might come from forming a coalition is elusive, it can still keep assets out of the hands of foes. Although this is clearly so for direct competition (such as war or politics), it holds even for indirect competition (such as business). A company developing software for one platform has that much less energy to devote to another platform. One that has adopted a standard for its internal network is unlikely to buy much that conforms to another one lest it create integration problems for itself.

Both demand and supply affect each member's decision to join a coalition. The demand factors are familiar. One joins a particular side based on how much one likes its members or dislikes their opponents and whether its victory, success, or ascendance is deemed helpful. The supply factors, less obvious, refer to the relative cost of working with coalition partners. The lower the costs, the greater the incentive to join. But the cost factors have many interesting hidden aspects.

6.2.2 Information and Coalitions

The greater the role of information in every phase of life, the greater a role that information exchange is likely to play in the economics of coalition formation and maintenance. Business coalitions can trade access to product design information (for example, computer-aided design [CAD] drawings), sales data, financial information, contact files, and inventory data.[7] In the political arena, contact files and mailing lists may be usefully traded. Military information that may be worth swapping ranges from intelligence to detailed situational awareness and incidents reports. Much as atomic nuclei are bound together by exchanges of quarks, so may coalitions be bound by exchanges of information in cyberspace. Such exchanges come in three forms: an exchange of information as such, access to each other's services, and connectivity with each other's members or external contacts.

What is it about information that gives it a special place on coalition formation?[8]

First, information is free in that its duplication and distribution are nearly costless – even if such largesse comes at the expense of potential lost sales of information or information services. In economic terms, its marginal cost is near zero. Its owners can give it away with fewer economic constraints than it could have for other things costly to duplicate or distribute. It is more readily brought to the table.

Second, the cost of receiving information and information services through cyberspace is *not* free or even, in some cases, cheap. Recipients of information must often invest in hardware, software, training, and doctrine to maximize the value of the information (or access, services, and such) they are awarded. Buying into a key E-commerce relationship

---

[7] Referring to a conversation with Monsanto's chair, Robert Shapiro, Friedman concluded:

> Monsanto . . . develops sophisticated new seed varieties that it markets to farmers, but needs to work closely with the big agriculture companies, such as Cargill, to make sure that Cargill will recognize the uniqueness of these new Monsanto-made varieties and then assign a higher value to crops grown with these new seeds so that farmers around the world will have an incentive to use these new seeds and pay more for them. Cargill and Monsanto have to know exactly what each is doing globally for Monsanto to reap the rewards of its scientific breakthroughs (Friedman, op. cit., 187).

[8] Some may object that what is called a "coalition" is really a market. This is true insofar as such exchanges can be settled in money. The intent, however, is to look at a broader range of exchange-based coalitions to include those among nations, militaries, political or professional groups, and nonprofit organizations where cash settlements play little or no role. Even commercial coalitions involve intangibles such as trust or common philosophy that are not fungible into cash.

may not yield much until the interfaces of one's systems change to work with another system and these, in turn, echo back to changes in internal functions such as order-processing and finance systems. Once the internals change, there is the possibility that such changes will echo back into one's production management systems and, perhaps even further back, to one's own suppliers. Owners of old systems may have to scrap what they own and retrain people and rewrite procedures to do things in new ways. Once there is enough invested in a relationship, a member may stick to it even well after the terms of trade have worsened and the original decision proves questionable. The ability to work with a product or system depends not only on its interfaces but also on what the users can expect from it and thus how deeply they orient their habits to their expectations of how it can function and in which circumstances. To exploit SAP AG's enterprise management software, for instance, requires serious users to redesign and often rethink how they account for time, material, and other resources.[9] This is a matter not simply of interface standards but of how information is understood. Until one such philosophy in that field become universal, transparent, and public, the adoption of open information standards, while helpful, will not eliminate the need to learn how to organize information as one's partners do. Many firms have already outsourced, or are looking hard to outsource, information services; that suggests that they are willing to revisit how certain parts of their business are conceptualized. Similarly, they may be willing to let their own systems echo the cognitive structures of their service providers.

Third, exchanges of information and information services, even if done *through* cyberspace, are trending toward exchanges *of* cyberspace.

---

[9] This was the experience of Jack Dangermond, CEO of ESRI, as relayed to the author. As described in a *Business Week* article:

> [SAP's] R/3 is a complex set of programs that can take several years and millions of dollars to roll out. It requires far-flung outposts of a company to adhere to the same, precise, business processes – forcing a company-wide re-engineering. "Implementing this type of software is not a technological exercise, it's an organization revolution," [says] Michael Hammer. Installing R/3 often involves an army of consultants that can cost three to five times the software's price tag. A $20 billion industry has grown up around SAP comprising consultants, trainers, specialists and hundreds of software makers who sell add-on programs ("Silicon Valley on the Rhine," *Business Week*, November 3, 1997, p. 164).

Information and information services can be exchanged in one of three ways:

- By transferring information directly, either on hard copy (for example, DVD-ROM) or electronically. Whether to e-mail a package or to stick it on a Web site for immediate downloading is less important. Either will do, and perhaps at the same time (for example, products could be delivered via downloading, and standards for using them could be available online).
- By creating a specific place (such as an external Web site) separate from one's internal cyberspace. Partners would have to be able to log into this Web site and interact with it, perhaps even with nearly as many privileges as those who set it up enjoy.
- By giving partners access into one's own portion of cyberspace.

The rise of the Web makes it easier and more cost-effective to give users access to information than to send it to them directly. Once there are enough transactions to pay up-front costs, hands-off exchange costs less. Hence the shift from information technology embodied as products (such as calendar software on CD-ROM) to Web-based services (such as calendar services). Microsoft's customer relationships used to be built on the presumption that once customers started putting Microsoft products in their enterprise architecture, they would stay loyal and purchase upgrades and upstream products. As Microsoft embraces the Web, it now offers periodic updates of its software automatically.[10] Its ".Net" initiative (described later in this chapter) was an attempt, in part, to shift from packaged software to a steady stream of software services. Similar trends are reflected in the desire of many start-ups and established companies to become over-the-Web application service providers (ASPs).[11]

---

[10] Or at least Microsoft is trying to do so. See Joe Wilcox, "Microsoft Program Meets Some Resistance," CNET News.com, http://news.com.com/2100-1001_3-908773.html, May 10, 2002.

[11] By mid-2000, *Business Week* found:

Roughly 4000 ASPs have been created in the past four years, offering "apps-on-tap" that run the gamut from corporate computing to the jazziest consumer services. Many have been scaled back, in the dot-com bust, with some conspicuous exceptions such as Salesforce.com ("Information Technology: Annual Report," *Business Week*, June 19, 2000, p. 73).

Should knowledge engineering, expert systems, or sophisticated natural-language processing prove technically feasible and useful, they may be easier to offer as a Web-accessible service than as software. Keeping knowledge current by updating one's own material and letting others access it compares favorably to figuring out periodically what has changed since the last update and getting it to users so that they can integrate new and old together.

The choice between setting up a welcome tent outside the enterprise perimeter and letting the guests into one's house depends, in large part, on one's relationship with them: do you trust them, can you control them once inside, and, most importantly, how closely do you want to deal with them? At the current state of Web security, it is relatively easy to use common gateway interface (CGI) scripts to hack and deface Web sites – and possibly go beyond the screen into associated files.[12] Security still matters.[13]

Unfortunately, frequent interaction creates a real burden of ensuring real-time synchronization between a member's internal system state (that is, information and applications) and the external sandbox. Suppliers, for instance, have to be apprised of one's inventories, as well as manufacturing and shipping schedules.[14] If such suppliers, for instance, are to fit their parts into the greater whole of their customer product, then vigorous electronic trading of computer-aided design and manufacturing (CAD/CAM) files, for instance, may be required. The rationale for using an external sandbox is that although external systems tend to be vulnerable, one can secure ground truth within one's internal systems

---

[12] Indeed, it is so easy that hacker wars have begun to focus primarily on such exchanges. In the spring of 2001 (following the crisis over the U.S. Navy's EP-3 that had to land in the Hainan province of China), attackers defaced two Chinese Web sites with slurs, insults, and even threats of nuclear war. During the following week, American hackers hit dozens more Chinese sites. Those supporting China responded by disfiguring one obscure U.S. Navy site. Carolyn Meinel, "Code Red for the Web," *Scientific American*, October 2001, pp. 42–51.

[13] Tales of credit card numbers stolen from hackers who entered corporate databases through Web sites are legion. For instance, 350,000 credit card numbers were stolen in January 2000 from CD Universe, an online music company. See Greg Sandoval, "Why Hackers Escape," http://news.com.com/2009-1017-912708.html, May 14, 2002.

[14] As John Fielder explained, "I can pull up on my computer right now Ford's inventories at all their plants because we have live data coming to us. We have less than a week's inventory of anything in our company." Interview in *Business Week*, May 20, 2002, p. 28D.

and use trustworthy internal information to correct corrupted external information, as Chapter 3 described. Security for the sandbox cannot be overlooked, though. While the sandbox is corrupted, partners will have been deceived and what they draw from it may be deemed untrustworthy.

Corporations are already forming extranets with trusted suppliers and customers, while using security controls to keep partners from sensitive files (for example, negotiations that one may be conducting with other competing partners).[15] Webcor, a San Francisco construction company,[16] for instance, has used its extranet to coordinate suppliers; on one project, it claims to have saved two months' time and a third of its project management costs. While multilevel security is still a hard problem, if one can select trustworthy partners, one need only worry about a few potential abusers who are also subject to more meaningful sanctions (such as loss of business) if their attempts to breach internal walls are discovered.

As a result, although coalitions within cyberspace – whether within a shared dedicated space or the space of one or another partner – are rare today, if and as security issues are resolved, they may become the sine qua non of coalitions tomorrow.

6.2.3 The Cost of Coalitions in Cyberspace

A core owner of a coalition is in a good position to manipulate the cost of entering or leaving a coalition in ways redolent of the underground drug trade. Entry costs can be low; only later are partners entrapped by webs of dependence on another's cyberspace – the information it contains, the contacts it enables, and the services it provides. The core partner may leverage this dependence to keep other partners on board. Depending on others for services is more binding than depending on

---

[15] PolyOne, a $3 billion polymer services company based in Cleveland, Ohio, recently integrated its SAP system with eight of its key partners' infrastructures to improve its forecasting, supply-chain visibility, optimization routines, and planning. For smaller partners, it used online exchanges. Tom Sullivan, "Take Your Medicine," *Infoworld*, August 31, 2001, p. 32.

[16] "Webcor ... adopted the habit of creating a special Web site for each project. All plans and timetables are laid out for employees, subcontractors and architects. If something changes, everybody knows it instantly." Marcia Stepanik, "Are You Web Smart?" *Business Week E-Biz*, January 18, 2000, p. 38.

information – especially for those who lack the means or confidence to determine which internal processes rely on such services to run at all.[17] There are also darker fears that the full dependence that pervades one's internal systems may leave one open for manipulation[18] (for more, see Chapter 9). The source of such vulnerability could range from one partner's general knowledge of the infrastructure owner's security doctrine, to specific knowledge of how the infrastructure is secured, to privileged access to the infrastructure that can permit an attack to be bootstrapped more easily. Known vulnerabilities might be fixed straightforwardly; unknown vulnerabilities, far less confidently.

One reason why dependence is more of an issue for coalitions that exchange information than it is to those that share other resources is that interactions in cyberspace are complex, and growing more so. Considerable investment is required to cope with the complexity, and the amount of complexity tends to rise steadily as the level of interaction among members deepens. Worse, this complexity is not solely external, but can ripple within the internal computer systems of each member of the coalition.

Standards issues reflect this complexity. Every information infrastructure supports conventions, procedures, and syntax that must be mastered to make it work at all. Infrastructures with open conventions, procedures, and syntax are easier to work with, largely because such standards are embedded in popular software; that is, someone else has already solved the interface problem. But the more specialized the domain or the deeper the interaction, the less often there is a common way of talking to it. The terms one employs to download a Web page with information on logistics may be ubiquitous and hence their use may be trivial or invisible (for example, the words are embedded in mouse clicks). Conversely, the

---

[17] Although packaged software can lead to dependence (for example, if one needs upgrades), software supplied as a continued stream of constantly updated services makes it difficult to know exactly the nature of the dependence. It may be harder to uninstall it because its installation has taken place over time and the various pieces may be finely distributed.

[18] Dependence and vulnerability may be related insofar as one side's operations require access to another side's services. If withdrawn, the partner's processes may fail in ways that create poorly handled software exceptions and thus security vulnerabilities – much as a wound creates the risk of sepsis.

vocabulary for putting logistics information into a scheduling simulation may be quite long, take time to master, and require reorganizing the logistics data to match. Sometimes the lack of a widespread standard arises because the domain, being highly specialized, lacks a critical mass to energize standards activity. But it could equally arise because too few vendors support such interactions, and none of them has much incentive to make its interfaces public or even easy to understand. In either case, coalition partners may therefore have to adopt certain standards, protocols, and conventions simply to interact effectively; these are not costless activities.

An extended relationship by users with one side's infrastructure tends to draw them into one's orbit. They have more working relationships both with the system itself and with others who use it. The categories that are used to process information specific to the coalition may become the categories used to process similar information even if it never sees a coalition facility or never goes to a coalition member. After all, why learn two ways of doing the same thing? If manipulating such vocabulary is analytically complex (for example, the logic by which the vocabulary is manipulated is unique to the vocabulary), it may well shape the perception of problems that the vocabulary addresses – something Chapter 10 describes in more detail.

The medical profession, for instance, has an extensive vocabulary to characterize illness and its treatment. It follows from the profession's tendency to instrumentalize work by consideration of the patient as a case – but the vocabulary reinforces that viewpoint as well.

The better that categories – where a language chooses to mark distinctions as significant – reflect how information is semantically structured, the more they tell about how people actually see things. This has always been true of human speech. However, putting these distinctions into software means that the manipulation of information is expressed in terms of rules (for example, if A is true, then B is true) executed by machines with little capacity for self-reflection or the ability to put such information in a broader linguistic context as people routinely do. If those who code the logic by which problems are analyzed adopt these semantic structures, they have to struggle not to adopt the rules that go with it. One might change the rules as such (for example, if A is true, then B rather than C is true), but it is harder to change the structure of rules and categorizations (for example, the definition of A, B, and C do not change) without alterations of the entire logic. Both anatomy and physiology, for instance, are different ways of dividing the body into subsystems and classifying parts of the body as members of one or another subsystem – but these subsystems interact differently at the anatomic and physiological levels. Overall, it is easier to buy into any particular logic among parts, subsystems, and the whole when things are simple, as they are more likely to be when relationships are formed. It is harder to back out when things are complex and pervasive.

The course of dependence will likely be affected by which partner has a better idea of what today's relationship portends for tomorrow's dependence. One *can* join coalitions, keep one's distance, and emerge with one's freedom of action intact. The cost of putting all relationships on an arm's-length basis in cyberspace will reflect, in large part, the evolution of commercial trends. Among other things, the more intimately that a core partner must know and may alter the details of another's information infrastructure and its state parameters (e.g., how the rules are set at any one point in time), the greater the potential dependence.[19]

---

[19] A good deal depends on exactly what the relationship is between the supplier and the buyer – lock-in may be the main point, not a side-effect. But see Michael Porter:

> Some observers argued that the Internet would raise [switching costs] substantially. A buyer would grow familiar with one company's user interface and would not want to bear the cost of finding, registering with, and learning to use a competitor's site, or in the case of industrial competitors, integrating a competitor's systems with its own. Moreover, since Internet commerce allows a company to accumulate knowledge of customers' buying behavior, the company would be able to provide more tailored offerings, better service, and greater purchasing convenience, all of which buyers would be loath to forfeit.... In reality...buyers can often switch suppliers with just a few mouse clicks and new technologies are systematically reducing switching costs even further...it is not enough for network effects to be present; to provide barriers to entry they also have to be proprietary to one company (Michael Porter, "Strategy and the Internet," *Harvard Business Review*, March 2001, pp. 63–78).

> McKinsey, a consulting company, reiterates the point: "Unlike open B2B marketplaces (orchestrated by third parties) and industry consortia (jointly owned by groups of buyers, suppliers, or both) private exchanges keep control in the hands of an active participant – an arrangement that helps focus activity on process rather than price." "The Unexpected Return of B2B," *McKinsey Quarterly*, http://news.com.com, www.mckinseyquarterly.com/article_abstract.aspx?ar=1210, September 1, 2002.

The logic of dependence need not favor the core partner. Economists Hal Varian and Carl Shapiro[20] argue that when the advantages of having others depend on you are clearly profitable, suppliers will compete among themselves to win customers, often by underpricing what they offer or even giving it away.[21] Circa 1999, this pattern was common among dotcoms, many of whom could then never figure out how to convert a free relationship into a paying one. Rapid changes in technology, economics, markets, or laws can break a relationship that is based on investments in having to install, train people to, and maintain systems based on existing standards. As circumstances change, sunk costs have to be written off and the game begins anew. Those who bid higher for partners may wind up with little pull on customers who can go elsewhere. If the latter are a large share of the business, the supplier may thus have little to show for its efforts.

The understandable wariness to enter into relationships that threaten to result in dependence is reflected in the continual battle over standards. As cyberspace has expanded, and the lessons of the personal computer era

---

[20] Carl Shapiro and Hal Varian, *Information Rules*, Boston (Harvard Business School Press), 1999. See especially Chapter 5, "Recognizing Lock-In." For the seminal arguments on the presence of lock-in, see W. Brian Arthur, "Positive Feedbacks in the Economy," *Scientific American*, February 1990, pp. 92–9; Charles Morris and Charles Ferguson, "How Architecture Wins Technology Wars," *Harvard Business Review*, March–April 1993, pp. 86–96. Conversely, Stan Liebowitz argues that claims of path dependence are weaker than popularly believed. It is theoretically impossible and, in practice, very difficult to find instances in which inferior technologies were chosen and locked into because of small rends in the fabric of prior history. Much of the theoretical argument requires that the owners of the inferior, but worse-faring, technology have deep pockets and a tolerance for throwing large sums of good money after if not bad then at least ill-fated money. Stan Liebowitz, "Should Technology Be a Concern of Antitrust Policy," *Harvard Journal of Law and Technology*, Summer 1996, pp. 283–318. See also idem, *Re-Thinking the Network Economy: The True Forces That Drive the Marketplace*, New York (AMACOM), 2002.

[21] The potential for lock-in following free services is widely recognized:

One popular conspiracy theory holds that sites are already preying on the addiction of their followers, hooking them on free services only to start charging when they seem indispensable. Although that may be true in some cases, many large Web companies would rather keep their sites free – they just haven't figured out how to make money that way.... "It's an old old model that venture capitalists used to use," said Clay Shirky, a partner at VC firm, the Accelerated Group, and a veteran Internet analyst, "Let them onto the front porch for free, and then charge them for coming into the house" (Stephanie Olsen, Jim Hu, and Mike Yamamoto, "Gated Communities on the Horizon," http://news.com.com/2009-1023_3-874724.html, June 4, 2001)

sink in, customers have become more vocal about avoiding anything proprietary. Those who buy a large share of a vendor's offerings (for example, a regional bell operating company [RBOC] buying phone switches) may have less worry – they can influence vendors directly. The rest may have to take pains to ensure that the critical interfaces between their world and a vendor's product – or tomorrow, its cyberspace – are transparent, public, tractable, and neutral. Vendors tend to be of two faces about open standards. Completely closed systems (such as those whose customers must rely on vendors for any additions or modifications) flash a red light to customers; they have largely gone out of fashion. Today's buyers are more comfortable if they know that third parties exist to contribute add-ons and support – even if the original vendor is the most likely supplier. Sometimes generally dominant vendors entering a specific market they have yet to dominate will embrace standards wholeheartedly – until they achieve the dominance they seek. Sooner or later, they will find themselves unable to resist extending the standard and giving customers one or another optional feature – a gift, as it were. This extension, in time, becomes an almost mandatory characteristic utilized by the other software. It may even be used by software that the customer itself writes. As a result, third-party applications written to open standards become, if not incompatible, then at least crippled in their usefulness and their ability to interact very richly with the rest of the system. Customers are thereby locked in. The logic that applies to vendors could also apply to owners of dominant cyberspace architectures, although it has yet to do so very much at this point.

If all parties to the coalition enter with a comparably clear-eyed view of where the relationship and the underlying technologies would take them, many of these machinations would be beside the point. There is an art and science of forming and exploiting coalitions, particularly among heterogeneous members – combining those who offer capabilities and those who accept and perhaps pay for them. A well-developed subfield of economics deals with bargaining power, negotiations, and game theory when one side knows something the other does not.[22] Yet, coalitions

---

[22] In 2001, the Nobel Prize committee paid tribute to the importance of understanding asymmetric information in negotiations by recognizing the work of Joseph Stiglitz, George Akerlof, and Michael Spence. Although most examples in the literature deal with

formed around information systems present two sources of asymmetry. First, information system owners tend to know details of their own systems (such as capabilities, offerings, and plans) better than users do – even if the relationship is open and aboveboard. Second, information technology relationships have, at least historically, been formed when there is little on the line. Those who participated had only a vague idea of its promises and potential (believing, for example, that personal computers were originally for geeks) and little serious money changed hands. As the technology matured and as systems grew more pervasive, the weight of the relationship began to matter. With that kind of prior history for personal computers and networks, it pays to be farsighted or at least suspicious when entering new relationships. That means knowing not only how important the linkage is to the other side and what aspects of it foster dependence, but how long it will last or how much maintenance it needs in its current form. Best yet, it helps to understand how the relationship will, in turn, transform one's partners.[23]

6.3 Enterprise Architectures and Influence

Information systems were originally built or bought to scratch this or that itch: control some manufacturing process, keep lists of customers and what they buy, or track money. Then they became ubiquitous, integrated, and hence intertwined – and why not? Shouldn't a customer purchase prompt changes in manufacturing schedules and budget forecasts? But, too often, information systems owned by their disparate departments could not talk to one another – or if they could, never figured out what they should communicate to one another. Hence the drive for systems integration. The commercial world flies flags labeled enterprise resource planning (ERP), customer relationship management (CRM), and others. DoD is building its Global Information Grid (GIG).

---

information about the product, the extension to information about the information system itself is a small one.

[23] By way of analogy, a central but rarely emphasized tenet of U.S. foreign policy is the hope that a working relationship with other countries permits our media (or at least our media culture) to pervade theirs, spreading our values and thereby democratizing them (for example, changing their internal power relationships). Some argue that all these shared norms would bind them more closely to the United States.

The challenge becomes one of carrying out integration thoughtfully, rather than simply having some vendor tell customers what they need (for example, what the vendor sells). Serious people thus must think thoroughly about the role that information plays in the enterprise, how it moves, and what it undergoes in transit.

To the extent that coalitions *through* cyberspace are coalitions *in* cyberspace – be they stand-alone or embedded – then the architecture of such enterprises is likely to affect (1) the efficiency of such a coalition as measured, in large part, by how easily information recipients work with information suppliers and (2) its ability to give the core members leverage over others.

Architecture usually refers to how the parts of a system relate to each other and the whole: the arrangement of materials in a room, the arrangement of rooms in a building, the arrangement of buildings in a neighborhood and a city. DoD canon refers to multiple architectures: a systems architecture (how components and/or nodes are linked in real space), an operational architecture (who talks to whom), and a technical architecture (how do components interface). One could easily add a data model architecture (the relationship of attributes in a database or among tag sets) or a security architecture.

Architecture may also be defined in terms of the rules, be they in software or legal code that influences (that is, prohibits, constrains, or encourages) how users or their agents, processes, systems, and so on access a system's resources, such as information, services, or connections.

Why should the architecture of an enterprise – how information is managed *internally* – matter to *external* relationships? The simplest answer is that inner form determines the shape of the interfaces. By analogy, take proteins. All proteins are built from the combination of amino acids, of which there are twenty basic types. Their shape results from which amino acids it is composed of and in what order each is linked to one another.[24] Since most proteins are long, complex, and heavily folded, only a fraction of their mass sits on the surface. The rest

---

[24] The macro shape of a protein is often determined by what links are formed among cysteine amino acids. The cysteine amino acid contains a sulfur atom, and thus two cysteine amino acids can attach to one another through bonds between the sulfur atoms on each.

is wrapped inside. Nevertheless, how it interacts with other molecules depends almost entirely on the shape of this surface. For a complex protein, therefore, architecture implies interface.

Enterprises that are, in effect, coalitions of their own employees will find that a good internal architecture facilitates efficient information exchange in ways parallel to how chemical catalysts predispose certain chemical reactions and not others. In some circumstances, architecture can express an enterprise's real command-and-control precepts.[25] For a complex enterprise architecture, system features that were originally designed for reasons of user satisfaction, cost efficiency, system administration convenience, and so on, are usually the features that others see and interact with. That very architecture is what influences how well its owner can forge productive alliances with outsiders and their information systems.

Certain facets of an enterprise's information architecture may predispose it to play a more useful role in forming and extracting value from coalitions over and above the content value of the information itself. If you want others to play and bring their toys, then you have to pay attention to how the sandbox works.[26]

Networking, security, and interoperability decisions, for example, affect a user's ability to access a system at all. They give some partners greater participation and influence than others.

- Networking – bandwidth, uptime, quality of service (QOS), and so on – is influenced by policy, protocols, and equipment. Because of the ubiquity of fiber optics, bandwidth is essentially unlimited

---

[25] Under its Force XXI architecture, the Army envisioned T1-like (1.5 megabits per second) connections into a battalion tactical operations center (TOC) and 9,600 baud connections from the battalion TOC to the field. Since a T1 line link should suffice to access all the information (except live video) that an Army warfighter needs, and no one higher in rank is closer to the field, effective command (determined not only by authority but by real-time knowledge) in the U.S. Army pivots at the level of battalion commander. Other Services (such as the Marines) or countries may pivot elsewhere. The Army explains the Task Force XXI concepts in its CD *Warfighters Digital Information Resource Guide*, December 1996.

[26] For a longer treatment, see this author's "Deconstructing Information Architectures," in Stuart Johnson, Martin Libicki, and Gregrory Treverton, ed., *New Challenges, New Techniques*, Santa Monica (RAND), 2003, pp. 67–95.

to most enterprises, at least in developed countries, but it may constrain those who sit in remote locations and individuals who connect through mobile devices (for example, most fielded military operations). Enterprises may wish to limit connection speeds either in hardware or via access policies in order to regulate other parameters (such as the risk of flooding from untrusted sources or from malformed database queries).

- Security, besides protecting cyberspace from the untrustworthy, also regulates who among the trusted can write and read what. The temptation exists for security architectures to be overelaborated, given the tendency to compartmentation (especially in government), the greater readiness of administrators to blame people for revealing rather than hoarding information, and the inevitable crankiness of the security authentication machinery.

- Interoperability problems can constrain information flow if systems cannot connect reliably – for example, if applications do not work together and data cannot be transferred without being stripped of some of their content or context. Rarely do owners deliberately raise interoperability barriers to sharing information (vendors are a different story). Yet barriers can persist because of the presence of legacy systems, the cost penalty from using general-purpose conventions, and the sheer complexity of ensuring that everything on both sides of a transaction aligns.[27] Sometimes, differences in the expression of material reflect underlying and deeper differences in the conceptual models upon which material is based. The eXtensible Markup Language (XML) has emerged as a protocol through which data may be tagged and categorized. But using it requires that everyone use the same tag sets – a process that has just begun.

Architecture also requires setting forth rules on the collection, organization, distribution, and presentation of information, or analytic tools for processing it. Some enterprises understand a central registry of critical and commonly understood data to be key to integration. Medical

---

[27] For instance, service delays resulting from the railroad merger between Union Pacific and Southern Pacific were exacerbated by difficulties in integrating their computer systems. "A Desperate Effort to Clear the Tracks," *Business Week*, March 2, 1998, p. 46.

information systems revolve around the treatment of patient records. High-intensity warfighting would focus on "what's where." E-business firms may adopt a transaction-centric model. Law enforcement may do the same for a crime-centric model. Combined, these facets influence how information is transmitted and expressed; they should reflect how the system's owners see the world. They set the terms on which information is gathered from others as well.

- Collection policy deals with what information enters a system (and when): as its quality, pedigree, and timeliness as well as where it came from, and what and whose questions it was designed to answer. A military information system, for instance, can be run for intelligence analysts or combatants. Analysts would scan for indicative events or anomalies against a background arrayed for consistency so that what needs attention will easily stand out. Combatants are more likely to generate information requirements afresh based on operational needs. Information will be provided faster and will fit requirements more tightly but will more likely be ad hoc and provide less basis for comparing today's conditions to yesterday's or establishing patterns against which anomalies can be detected. Third parties may have an easier time plugging into the top-down systems built for intelligence operators since such systems have a more stable background; information systems for combatants must continually adapt to changing requirements.
- Distribution policy influences how information is circulated. Pull methods include posted files (for example, Web pages) both generalized and personalized (such as how Amazon.com's first page to a viewer is based on recent purchases), responses to queries (to a database), and more complex variants (such as a guided tour whose paths are based on prior choices). Push methods include e-mail, instant messaging, alerts, and monitors as well as semiactive methods such as tailored newsfeeds. Comparing the last item from each list suggests that interaction has elements of both push and pull.
- The organization of information encompasses both its classification schemes (such as the Library of Congress method) and the tools to find things. Today's search tools keep getting better but they are neither perfect nor neutral; they can be and are manipulated.

Any one system of organizing information helps put certain pieces of it in ready context with one another; compare a map and a gazetteer with full latitude and longitude information. Classification metaphors (for example, dictionaries, histories, manuals, and newspapers) influence what people see and how they see it – and whose questions are more easily and reliably answered by the system.

- Presentation also influences perception, despite the hype that all information can ultimately be racked, stacked, sliced, and diced to any user's unique perspective. In practice, certainly at DoD, the presentation is often of fixed format; for example, DoD's Common Operational Picture of the battlefield was established without the macro languages that users might otherwise have used to tweak the picture to their needs. Perceptions tend to follow. As justification, designers cite lower costs for training and user support, guaranteed interoperability between information and application, and the need to minimize perceptual divergence between those who must work together.

- Analytic tools convert certain forms of knowledge (such as images) into others (such as maps). In most cases, conversion is straightforward. Depending on the progress of knowledge engineering, systems may be capable of coming to certain conclusions based on the application of logic rules to data.

Although architectures are designed for efficient information exchange, they can also be used to attract coalition partners, extract value from their membership, shape the terms of trade with them, and raise barriers to their departure.[28] Efficiency may be secondary when setting usage policies. Such policies may instead be framed to induce others

---

[28] Those who would shape their architecture to build better coalitions should remember that, if they are not careful, such decisions might also color how *insiders* interact with the system. Features such as collection, distribution, organization, presentation, and analytic tools tend to look the same from inside as from outside (imagine the difficulties if a system filed data one way for internal users and another way for external ones). External connectivity, security, and interoperability rules, however, may well look different when viewed from the outside compared to the inside (for example, modem connections outside versus Ethernet inside). Yet, insofar as large enterprises (such as DoD) divide their internal world into multiple compartments – rather than simply differentiate what is inside from what is outside – external constraints end up becoming internal ones.

to tender valuable information (for commercial Web sites, for example, such information may be demographic data or personal shopping preferences), to conform to interoperability conventions, or to support certain perspectives with the information they present. These goals overlap. The more value a partner gets from participating in another's cyberspace, the greater is the efficiency of the partnership and the fewer the inducements to defect.

Cyberspace is still young. There is much to be learned about what sort of manipulation may prove particularly useful – as well as what sort of techniques partners on the receiving end may employ to shield themselves from manipulation.

6.4 Alliances with Individuals

The next two sections should convey a sense of when and how to exercise influence via cyberspace. This one looks at alliances made with individuals. The next looks at those among enterprises.

Is cyberspace a medium by which individual loyalties can be won and mobilized? Subtle but real distinctions exist between the quality of influence available through mass media and those associated with one-to-one media such as the Internet – even if wielding influence by feeding CNN or by feeding cnn.com are not so different.

One-to-one media, like mass media, can reach many people quickly and mobilize them for action. The middle-class anticorruption uprisings in Bangkok (1993), Djakarta (1998), and Manila (2000) were organized via one-to-one cell phone usage; the silent protest of fifteen thousand Fulangong members before the central leadership compound in Beijing of April 1999 was organized via one-to-one e-mail. Accidents along well-traveled roads prompt a flurry of 911 calls; indeed, safety officials now worry that accidents are more likely to be overreported than underreported.

One-to-one media have several advantages over mass media.[29] Some mass media are simply unavailable for such uses, being in law or in

---

[29] See, for instance, Tiffany Danitz and Walter P. Strobel, "Networking Dissent," in John Arquilla and David Ronfeldt, ed., *Networks and Netwars,* Santa Monica (RAND), 2002, pp. 129–69.

practice[30] under government control. One-to-one media also permit activity to be organized without appearing on the radar screen of authorities or other opponents. Such activity can be organized more selectively by, for instance, not giving certain people the message. Location-based messaging in future generations of cell phones[31] may permit even more efficient public mobilization, being adaptable to circumstances on the street within minutes. Services such as instant messaging or GSM's short message service (SMS) may make it easier for news to be distributed person to person via interlinked circles of connected friends, thus breaking mass media's control over the dissemination of news. This does not inherently favor DoD (or, more generally, does not reward the sophistication of an enterprise's architecture), although DoD can learn to play.

One-to-one media enables users to tailor content to individuals, although tailoring it well is not trivial. Even a well-edited newspaper cannot help but deliver the same news to all, with the modest exceptions such as regional editions of metropolitan newspapers. In cyberspace, there is more opportunity for individualization based on (1) one's e-mail address; (2) one's physical location at the time, especially for those who reach the Web via mobile phones; (3) voluntarily provided customer profiles; (4) correlated information garnered about an individual that can be garnered for public and quasipublic files; (5) patterns of virtual interactions (for example, the clickstream); and (6) the history of real interactions, such as purchases. Advertisements, individual newsfeeds (such as those that provide information to a user requesting news about a particular baseball team or on current events in Tunisia), and affinity groups (such as links to a chat room a particular user may find interesting) may all benefit.

The information overload discussed in the last chapter provides more than enough reason to tailor information to the individual. True, those who would filter the news may impart their own biases and worldview to

---

[30] Italy and Thailand, both democracies, maintained the practice of having two major television networks, one owned by the government and the other privately. In recent years, however, the private television station owner ended up becoming prime minister (Silvio Berlusconi and Thaksin Shinawatra, respectively) and running the government. This effectively nullified what, on paper, had been diverse ownership.

[31] AT&T became the first U.S. cell phone provider to market location-based services ("Find Friends"). It used cell-based triangulation (rather than nascent Enhanced 911 capabilities) to locate someone within a block in well-served cities and within a few miles in rural areas.

their suggestions of what to see. But if such people were not like-minded and trustworthy, viewers, given a plethora of choices, would not have bought into their filtration and presentation services in the first place. Indeed, cyberspace lends itself to being organized as like-minded villages, a process that echoes earlier urbanization patterns. Historically, migrants to the big city from the countryside have tended to come for the money and put up with urban living as a necessary evil. Many reacted to the social disruption by creating unicultural neighborhoods that limited their exposure to a frightening multicultural world. By the same token, many people coming to cyberspace will have arrived for economic reasons and may take similar refuge among familiar virtual faces. But urban villages[32] are not villages; they do not last forever, and their residents – or at least their children – eventually assimilate. Similarly, reacting to the anarchy of cyberspace by finding a comfortable bolt hole will not forever retard the process by which people get comfortable with ubiquitous information from all over.

A service in cyberspace can exercise a sticky hold on users beyond that of the attractiveness of its content. It does so by persuading them to entrust it with something of value. One who wishes to switch service providers may find that too many people have to be informed of the address change – hence the link between cell phone number portability and competition.[33] Users of Web-based mail services often find that, unlike many client-side E-mail systems, the site rather than the user retains all the correspondence and users can lose access to it for any of many infractions including – if paid service ever becomes the norm – not paying one's bills. Those who enter scheduling information on a Web calendaring site may find they cannot transfer the information back to another product; they may therefore rely on the provider to the extent that such information is either valuable or hard to transfer back. This echoes how proprietary standards keep people from switching software. Yet, it remains to be seen how much value individuals are willing to leave behind in cyberspace.

---

[32] See Herbert J. Gans, *The Urban Villagers; Group and Class in the Life of Italian-Americans,* Glencoe (Free Press), 1962.

[33] See Ben Charny, "It's Your Number; Take It with You," http://news.com.com/2100-1037-5100892.html, November 3, 2003.

One-to-one services also permit the resources of individuals to be pooled. A benign case is the virtual supercomputer that the search for extraterrestrial intelligence (SETI)-at-home folks built by persuading millions of people to run their analytical program as a screen saver.[34] Distributed denial-of-service attacks are a malign case. Serious business models are based on the hope that idle corporate processing capacity can be harnessed to solve difficult problems.[35] Yet, it is unclear how many such problems can be broken down as finely as SETI-at-home, which divides the universe of received radio-frequency energy into specific signal blocks for discrete analysis.

6.4.1 The Special Case of Cell Phones

The number of people accessing the Internet via mobile phones may soon exceed the number doing so from wireline devices. Today's phones support some basic text messaging. Japan's NTT introduced iMode (as part of its DoCoMo service) several years ago and now boasts over 40 million users of slow-speed, low-resolution, but always-on Web services. Long-awaited 3G services promise bandwidth of roughly 384,000 bits per second.[36] This is nothing to sneeze at if it only has to drive a hand-held device (for example, a Palm PC such as the Compaq Ipaq with a full color 240 × 320 window or 180 × 200 screen on cell phones that display snapshots). This rate can support MP3 audio, fast snapshot transmission, fairly clean videotelephony, and the presentation of Web pages faster than they can be scrolled. Full-motion video at that resolution rarely needs more than a megabit per second. Wireless receivers are becoming an electronic version of a Swiss Army knife, combining not only cell

---

[34] By some estimates, 10 percent of all ostensible SETI screen savers are actually processing data from rogue servers.

[35] E.g., Stephen Shankland, "Energy Dept., IBM to Unveil Science Grid," http://news.com/2100-1001-866452.html, March 21, 2001. But Jeff Clarke takes a much more skeptical view of its economics. Jeff Clarke, "The Grail of Utility Computing," http://news.com/2102-7339_3-5091350.html, October 15, 2003.

[36] As of mid-2005, one vendor, Verizon Wireless, had expanded coverage of its 1xEV-DO 3G cellular data service to fifty-five U.S. airports and fifty-one metropolitan areas. Typical speeds are 300,000 to 500,000 bits per second for roughly $80 a month for unlimited access. See www.mobilepipeline.com/164903360.

phone and PDA functions, but also MP3-level audio, cameras, GPS, and remote control. Overseas, cell phones are used as credit card devices for vending machines. Power permitting, they could be continuously connected to the Internet. This possibility has already excited vendors with the prospects of M-commerce (*M* for mobile) based on customizing information, advertisements, coupons, and so on, to people based on their location.[37] One can only imagine what the crowded cities of third world countries will look and feel like when every other person is toting such a device, just as every second pedestrian in Europe seems to be talking on one. Events in the real world will be continuously conveyed into cyberspace, with some chunk of that virtual world being posted for continuous public consumption.

The dissemination of such technology throughout the world, in rich and not-so-rich countries alike, creates both enormous opportunities and challenges for national security. Video cameras are already ubiquitous in Iraq's cities, and are particularly prized by insurgents. To see the contrast, ponder a U.S. platoon slogging its way through a village during the Vietnam War. It could have been in, out, and 10 kilometers down the road before anyone in the village could make contact with the enemy and give away the platoon's range and bearing. The same trek through what may be more of a middle-income country could soon, perhaps now, be instantly captured by video, georeferenced to within 20 meters, and then posted to the Web for everyone everywhere to see.

Although the world is becoming more transparent, can transparency be fogged in a war zone? Cell phones and cellular signals are fragile and single-mode GPS can be jammed. Infrastructure can be destroyed and subscription services hard to collect money for; prepaid cards, though, pose fewer problems. Cell phone usage is easy to trace – by person if the call is unencrypted, by equipment even if it is, and by RF emissions

---

[37] Well, perhaps some day M-commerce will be possible. According to Troy Wolverton:

Customers [in the United States] are showing less enthusiasm for mobile commerce according to biannual surveys by A.T.Kearney. Consumers' intent to make purchases using cell phones fell by about two-thirds in the last two years..., 32 percent of all cell phone users planned to make purchases; by January 2002, that number was down to just 1 percent. A.T.Kearney surveyed 5,600 cell phone users in Asia, the United States and Europe (Troy Wolverton, "Mobile Commerce Rings Up No Sale," http://news.com.com/2100-1017-956969.html, September 9, 2002).

when law proscribes all transmitting devices. Yet, with every passing year, the prospect of squelching mobile telephony except locally and in emergencies grows less tenable. More people will be upset. More commerce will be disrupted; cell phones are how business gets done in some places.

The alternative strategy is to keep cell service going and exploit it. People who tote digital cell phones could very well become the eyes and ears of a watchful society. They could provide information, including still or moving imagery, on events as they happen. Such reports could have excellent geospatial accuracy from knowing where the phone is plus maybe some rough-and-ready range finding from the caller to the event and valuable voice annotation. In a military that prizes information superiority as key to victory, what better sensors could it employ than hundreds of thousands of visual and aural devices of the metropolis all connected through very intelligent processors?

But how could one get them on one's side?

The challenge for the U.S. military – or any security regime that would tap into high-tech people power – is in drawing in as much good information as possible while filtering out the bad. If today's IM is indicative, garbage can be expected even without malice.

Why would people call in? They may do so out of civic duty or because they favor one side. Anything that makes a contribution technically easier and quicker would help. Unfortunately, the Web is littered with ill-considered ideas to induce largely commercial participation in this or that site. Marketers that tried to induce "community" formation have not done famously well.[38] Consumers have rejected European phone manufacturers that tried to make it easy to link with favorite suppliers and hard to link with their competitors.[39]

---

[38] See, for example, Ellen Neuborne "How Commercial Communities Can Click," www.businessweek.com/technology/content/may2001/tc2001057.248.htm, May 2001. "Companies trying to get personal with their Web site visitors in hopes of increasing sales are wasting more money than they're earning warns an upcoming report." Paul Festa, "Beyond the Personalization Myth by Jupiter Research," http://news.com.com/2100-1038-5090716.html, October 13, 2003.

[39] "The special business tie-ups can make services frustratingly limited. Your operator may advertise access to movie schedules that turn out to be for only one chain of cinemas – other chains may have cut deals with competitors overseas" ("Breaking Down WAP's Walls," *Far Eastern Economic Review*, September 14, 2000, p. 36).

Then there's the risk of passing on tattle-tale reportage about people who carry grudges and weapons. Can the phone company be trusted not to report one's cell phone location – as U.S. cell phones are gaining the ability to report their locations? In the United States, where almost all cell phone calls are billed to a person, further identification is possible; however, overseas, prepaid phone cards – which offer anonymity – are more common. Trusted anonymizers (who relay calls and wipe out original source information) may reduce such fears, but trust in the anonymizer cannot be assumed[40] and callers must also trust that their cell phone is not corrupted to slip identifying bits into the stream. Foreigners may not trust the security of their own security services; they could be compromised, as many were in Vietnam and are in Iraq.[41] They may also be expected to have much less faith in foreign militaries (such as DoD).

Witnesses may resort to anonymous reporting, but doing so complicates the problem of filtering out bad information. Digitally signature algorithms (for example, those based on an embedded identifier plus a user-supplied PIN) may segregate as unreliable information from callers who are not identified as trusted, but trusted callers are likely to be, at best, a few among many. Other verification methods include waiting for multiply sourced reports (with some way to distinguish them from multiple single-source reports), polling passersby (for example, ringing everyone within some radius of the event) for some confirmation of events, or looking for a correlation between reportage and information from organic sources (such as one's forces or one's sensors).

---

[40] The Church of Scientology talked Finnish law enforcement authorities into forcing Johan Helsingius, the owner of a prominent anonymizer site (www.penet.fi), to yield the name of someone who had posted Church materials on Usenet by going through that site. For a discussion of this case and other issues related to anonymity over the Internet, see Robyn Wagner, "Don't Shoot the Messenger: Limiting the Liability of Anonymous Remailer Operations," 32 *New Mexico Law Review*, Rev 99 (the "Frontiers of Law: The Internet and Cyberspace" issue), pp. 99–142.

[41] In at least one case, someone who forwarded information on an ongoing attack to authorities was discovered and killed when his phone number was sold to insurgents. Lawrence Kaplan, "Centripetal Force: The Case for Staying in Iraq," *New Republic*, March 6, 2006, pp. 19–23.

6.5 Alliances of Organizations

As the following examples suggest, alliances in cyberspace offer considerable scope for both efficiency and relative power. The first deals with how technology develops through the give and take of challenges and responses; it then considers how sophisticated information systems can grease such exchanges and fuse various links on the supply chain. The second shows how DoD, which is creating an information system to illuminate the battlespace, can leverage access to such a system to glue military coalitions together. But with this comes responsibilities, especially in preexisting alliances, as laid out in the discussion of the NATO perspective, the third example.

6.5.1 Ecologies of Technological Development

Technological development (for example, computers, aerospace sectors, and even automobiles or financial instruments) requires a great deal of good information on the latest market requirements as well as the newest and best inputs (including supplies, tools, and venture capital). Such knowledge not only helps direct effort efficiently but makes one an attractive business partner.

Figure 6 shows how such relationships, which together form an ecosystem, may tie together prime contractors, subcontractors, and suppliers of services such as finance or accounting. In this example, the prime contractor has a problem requiring new knowledge to solve. The information required to pose the problem is complex, difficult to describe, and may not be fully understood unless one has worked with the vendor. Thus, the problem has to be farmed out to favored suppliers (for example, subcontractors or service suppliers). Receiving the problem is valuable for them. It indicates where to focus their development efforts – whose results may then be applied beyond the current problem or even the original requester. It also offers an implicit promise of business for those who solve the problem. This can work in reverse. A key supplier (for example, Intel) previews its products for specific clients, and by so doing presents an opportunity to be exploited rather than a problem to be solved. Similar flows may characterize training or financing opportunities. The growth

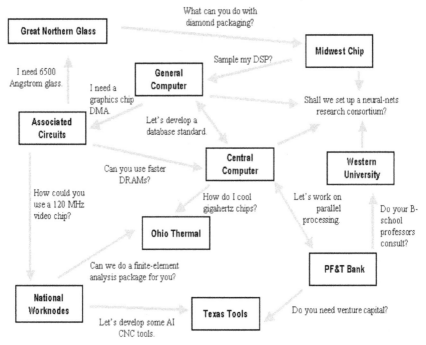

Figure 6. Production Relationships in Cyberspace
Note: This diagram is a slightly revised version of one that appears in the author's *What Makes Industries Strategic?* McNair Paper No. 5, Washington (National Defense University Press) November 1989, p. 9.

and continued prosperity of regional clusters (such as Silicon Valley, Wall Street, Hollywood, and Japan's automobile sectors) underlines this logic. Silicon Valley, for instance, is hardly the lowest-cost place to make silicon products. Once the designs are routinized and the procedures are down pat, places like Eugene, Boise, or Albuquerque have more economical plant sites. But as long as new information is key to innovation and as long as innovation is key to profits, Silicon Valley is the place where one can get more of the relevant information for less effort than anywhere else in the world.

The trend toward cultivating favored suppliers rather than having the broad market bid on them originated in Japan.[42] It spread to the

---

[42] That it started in Japan is partially due to the important role played by mutual obligation in that culture.

United States in the late 1980s and early 1990s as many large companies (including Ford, Xerox, and General Electric) took pride in how far they had thinned their supplier list. Japan envy aside, proliferation was spurred by the recognition that the information interface between buyer and seller had grown more complex. Products had become more sophisticated and there was less tolerance for deviation between what was asked for and what was delivered, in terms of both product specifications and contract fulfillment. More information had to be conveyed and understood. This information was difficult to transfer anonymously and remotely. It required a close working relationship with, in some cases, frequent person-to-person contact. Hence, it required alliances where each partner could devote time and attention to making such a relationship work; many equally valuable but less-directed interactions took place off hours among employees of various companies in Japan.

If recent trends toward electronic marketplaces reflect a pullback, it is because information technologies have lowered the cost of transmitting information; Japan has also lost some luster as a model. Companies can rebalance the trade-off between intensive information exchanges that require close relationships and extensive information exchanges that enable a wider range of suppliers to compete for business, thanks to the Web. Standard content description languages permit a richer exchange of information that machines can preprocess for humans (for example, by sorting potentially interesting material). Advanced enterprises are thus thinking about how to refer in standard ways to common concepts by using the XML standard and tag sets that are coming with it.[43]

More designs, plans, technology vectors, market appraisals, and such may come to be passed among prime and subcontractors via prime-subcontractor extranets. Proctor and Gamble (P&G), for example, gets detailed information on Wal-Mart's stock flows so that it can assume responsibility for managing the inventory of P&G products in Wal-Mart's hands; P&G then analyzes such subsequent sales data for its own

---

[43] According to the GAO, the federal government may be getting ahead of itself in trying to do so at this stage of its enterprise architecture; see Margeret Kane, "Government Seeks Accord on XML," http://news.com.com/2100-1001-935223.html, June 17, 2002.

information.[44] Five hundred companies have access to a Herman Miller system that permits them to check what factory's needs will be in coming weeks; if inventories are less than a day's worth, they promise to ship product.[45] Automobile companies can specify components by allowing potential subcontractors to interact with a three-dimensional model of the automobile.[46] Subcontractors' production and shipping schedules can then interact with the automaker's material-resources planning (MRP) program.

Although such interactions can be standardized (by using XML, for example) to make the same part description roughly compatible with every automaker's model, the degree of compatibility is likely to be limited for a long time to come. Some automakers may be interested in one particular feature of a component, others in another feature. The computer models that permit subcontractors to evaluate their designs, meld their production schedules, or conduct cost-quality trade-offs are likely to be proprietary in whole or in key parts (such as each automaker's design optimization routines). Each prime contractor will have to determine what can be revealed about its automobile – its key specifications, its critical components, or its manufacturing process and schedule – without yielding intellectual property. Only trusted supply firms are likely to be granted access to such a model. Conversely, only self-selected subcontractors will choose to invest the time and money to learn how to interact intelligently with such models. That alone would reinforce any

---

[44] Outside firms such as P&G employ whole teams of analysts in northwest Arkansas just to monitor the flow of goods that are shipped and sold through Wal-Mart. See walmart.nwanews.com/wm_story.php?paper=adg&storyid=121067, July 3, 2005.

[45] David Rocks, "Reinventing Herman Miller," *Business Week E-Biz*, April 3, 2000, p. 90.

[46] As just one example, note the description of Valeo's interactions with automakers: "Whatever it makes, however, it has to work closely with the car manufacturers that are its customers. The more complex the product, the closer must be the cooperation. The company is increasingly involved at the earliest design stages of a new vehicle." "E-Strategy Brief Valeo: Less than the Sum of Its Parts,"*Economist*, June 23, 2001, pp. 73–4. Daimler-Chrysler's FastCar project will given some four thousand internal and five thousand external users access to an interactive three-dimensional home page; even rental car companies would be written into the project. Alan Hall, "How the Web Is Retooling Detroit," *Business Week*, November 27, 2000, p. 194B. Moen, the faucet maker, launched ProjectNet, an online site where it can share digital designs simultaneously with suppliers worldwide, with all design changes consolidated into a master file in near real time. Faith Keenan, "Opening the Spigot," *Business Week E-Biz*, June 2001, p. 4.

tendency to rely on a few close relationships rather than many loose ones. Such relationships are not casual; each partner must trust the information it gets from another and must have sufficient prospects that serious sales will result from close participation. The ability of an organization to form such relationships in the future – and its information architecture will be a factor here – will therefore remain critical in making the right relationships at reasonable cost.

The challenge for those who run the relevant information forum (such as the carmaker) is how to design one that helps meet these goals. It might want working relationships that yield knowledge without requiring suppliers to pay unreasonable entry costs. Or, it might engineer relationships so that suppliers are more invested in the forum than the primes are – the better to exercise power over them.

6.5.2 DoD's Global Information Grid (GIG)

Cyberspace may be key to how the United States – and by extension, those it fights alongside – go to war.

Traditionally wars were fought by massing human or mechanical firepower and throwing it against enemy forces. With the development of precision guided munitions (PGMs), notably since the 1970s, the identification and exact location of a target suffices to put it at terminal peril. Taken to its logical conclusion, warfare becomes a matter of finding targets while not becoming one. When casualties must be minimized, finding is best done by using sensors to gather raw data on the battlespace and processors to convert the raw data into information. When this information is combined with one's own location, requirements, status, and plans, targets can be generated. Communications and display devices can then be used to display operational assignments against targets. Finally, precision weapons convert targets into ruins. For this reason, the tenet that modern warfighting requires a "system of systems" has passed beyond mere logic to faith. The emerging agglomeration of DoD's information systems is known as the Global Information Grid (GIG).[47]

---

[47] To be fair, DoD's GIG requires major investment at a time when shooters, rather than hunters and seekers, still run the armed services. Interoperability problems, especially among legacy defense systems, are still severe. Even if the various services

The more thoroughly each of these steps is integrated with each other, the faster the cycle can be executed, the fewer the people put at risk, and the rarer the mistakes. One such task, data fusion, requires processing nodes to receive sensor data in real time, battle management software to run these sensors, and battle damage assessment to task and retask operators. The agglomeration must be simultaneously tight to meet demanding operational parameters, yet flexible enough to accommodate the irritating habit of enemies to do the unexpected. Few civilian systems have such exacting real-time requirements or work in such deliberately unpredictable environments. This advanced feature of military cyberspace has the potential to drive the state of the art in systems integration – or disappoint its proponents if the state of the art cannot be so advanced.

The ability to transmit targeting data anywhere permits a commander to choose among any of several shooters to go after a target, thereby demoting the importance of any one shooter in the process of destroying it. The weapon could easily come from shooters too far away or too well hidden to be at risk themselves. Finders need not be killers. A transfer of processing from weapons ("here's where to look for a target; now go search") to external sensors ("here's where the target is within 10 meters; when you get there, scan a small circle around that location to determine the impact point") keeps weapons cheap and thus assignable to a wider class of targets.

The shift to distributed sensors, each feeding partial information to a central process, logically connected even if physically distributed, means that the center of gravity of a modern military would not be located with any one weapon. It would be in cyberspace – a meaning captured by the term "network-centric."

Separating information about operations from operations themselves gives DoD new tools to deter or affect the outcome of war. It need not deploy so much military force if it can leverage the contribution of allies

build their various systems (such as the Navy's Cooperative Engagement Capability [CEC]), their extension across service lines (much less international ones) lags (for example, the Army's Patriot missile has not played well with the Navy's CEC).

by letting them exploit information in the GIG. Here, too, cyberspace is the potential fulcrum for new power relationships.[48] It reflects the possibility that the information that the U.S. military uses for combat purposes could have a value independent of, and in some cases higher than, any ability of U.S. forces to convert such information into combat power.

In some circumstances, after all, it may simply take too much time to deploy U.S. forces to defend an ally. If such allies had their own munitions, it may be faster to supply the information – from cyberspace or at least through cyberspace – that can ensure that the munitions shot by allies hit exactly the right set of targets.[49] Compared to deploying forces, information transfer carries less military and even political risk. Information other than aimpoints can also be provided this way: intelligence and other background data (such as maps), maintenance instructions and algorithms, medical advice, training information, as well as software for data analysis and decision support.

Recipients, to be sure, must have a command-and-control culture that appreciates how to use such information. Although general intelligence on the enemy's whys and wherefores is always useful, exploiting precise targeting information requires precision weapons. The United States also has to be comfortable with how such information is used. The value to the U.S. military of plugging its allies into its cyberspace is realized only when they can use such information well. The ability to interoperate with the United States more smoothly cannot help but encourage a closer military-to-military relationship. The relationship in cyberspace between their military and DoD may even, in some circumstances, color decisions on whether other countries ally themselves operationally or pull back and create a more independent military capability.

---

[48] This point is made at greater length in a prior monograph of the author, *Illuminating Tomorrow's War*, McNair Paper 61, Washington (NDU Press), October 1999.

[49] The problem of acquiring this information at least begs the question of where the sensors are to sit all this time. Some of them can sit in space, but they will probably have to be complemented by UAV-borne sensors, ground-based sensors, and, in some locations, sea-based sensors – all of which have to be near or in the theater. So, something either has to be forward or go forward in crisis. Nevertheless, as the costs (and power requirements) of sensors and processors decline, it become easier to maintain eyes rather than arms in remote theaters.

How the GIG is organized will also affect what use friends can make of it. What, for instance, is the relationship between their initial investment in using the information and the value they get from it? What are users allowed to see and will they be allowed to see it when they want to? Can they see the raw data underlying the information? How exportable is the information and in what form? What information must users themselves reveal in order to see the more sensitive material? How easily can users find, access, and exploit information?

The degree of technical interoperability will also influence the gap between what DoD might offer and what others use. Will the information or the information services developed for U.S. operational requirements also answer the needs of friends? U.S. aircraft that tote radar-suppressing weapons may permit attack fighters to evade enemy fire by flying above the range of man-portable missiles. Another air force, without such suppression systems, may instead seek safety in flying fast and low. Both sides will need different threat information. Will the United States be willing or even able to collect, analyze, and present a different air threat picture for allies? Do the command-and-control rules reflected in construction of the GIG, notably in sensor management, comport with how other militaries work? A system built around satellites is more likely to construct its perspective from the top down accordingly to global priorities; it may also afford more control from the top. One built around tactical UAVs is more likely to reflect the ground-level exigencies and frustrate control from the top. One that is built around human observers is more likely to be variegated and inconsistent, but much less likely to leave anomalies unexplored for their not having been programmed in advance.

The terms of trade demand attention. What should allies expect to contribute in return for being able to tap into the GIG? Does it suffice that they be passive users? Should they give the United States explicit permission to log their inquiries – which also means their promising not to try to cover their tracks through DoD's cyberspace? Should their access be more interactive in that they are forced to give plausible reasons for wanting to access this or that part of cyberspace? Should they be asked to contribute their own reports or sensor data? If so, can and should the United States set down standards for reporting formats, frequency, coverage, and so on? If it does, is the United States wise to instrument its allies' sensors?

One of the implicit quid pro quos for giving other countries access to DoD information is that they format their own material in the same way before handing it over. This hardly seems controversial, but it could well alter the militaries that do so. At one level, sufficient exposure alone to such an informaticized approach to warfare is bound to change those who thereafter cannot go back to their old habits.[50] There is also an inevitable education of foreign warfighters that comes from their prolonged exposure to those categories of thought that reflect what the United States believes is important to know about the battlefield. Either or both can promote the evolution of the user's force structure to take the continued flow of U.S. information into account (for example, the aforementioned requirement for precision weaponry to exploit precise information). The change may be tacit; to interoperate with the United States requires a user to adopt technical standards that presume a certain architecture of information flow; examples include all-points versus hierarchical access, organization of knowledge by area rather than by weapons specialty, and a robust capability for annotation.

Exposure to a technical architecture could spread compatible information flow architectures. If DoD, for instance, expends effort gathering a geospatially referenced electronic order-of-battle and passes it to its friends, recipients might find ways to exploit such information in their battle planning, and even collect complementary data. Similarly, if DoD establishes a social network database (one that links various people to one another) and develops and distributes tools for maintaining and exploiting such a database, then intelligence collected by its allies may be analyzed for how it might populate such a database.

---

[50] This holds even more if training accompanies access to information. As a related example, the *Economist* observed "China has been more concerned about the political implications of Aegis. For Taiwan to use the system effectively, the US would need to train Taiwanese military personnel and integrate Taiwan into the US military-satellite network. Beijing would see this as the establishment of a quasi-alliance between Taiwan and the US" *Economist,* economist.com/agenda/displaystory.cfm?story_ie=E1_VJQSDR, April 27, 2001.

Will users spend more or less on acquiring information if they think they can tap into U.S. sources? According to what economists call substitution effects, expenditures will decrease. Conversely, exposure to good information may induce a user to realize its value, want more of it, and get more of it; bringing precision weapons to exploit precision information from the GIG establishes a greater demand for one's own precision information. Having users adapt themselves to U.S. systems even on a trial or infrequent basis (as in peace operations) flattens the learning curve of working with the United States on a more permanent basis later on. Irrespective of each country's motivation to ally themselves with the United States politically, the barriers to a *military* alignment, and thus to effective interoperation, can be reduced.

6.5.3 Merging the Infrastructures of Allies

At other times, owning a dominant infrastructure may seem all problem and no profit. In transforming its military around an information backbone, the United States has encountered backbiting from its NATO allies. A coalition that worked well when everyone had comparable, which is to say compatible, technology may, it is feared,[51] work poorly – and thus lose its cohesion – with the U.S. military so technologically dominant. The United States may exploit information technology to make fast decisions, maneuver stealthily, and strike precisely with limited collateral damage. Allies, not so equipped, would not only account for less of the total combat power, but could even retard decisions, produce unacceptable collateral damage during combined operations, and reveal allied positions and strategies by creating telltale signatures (such as noises and radar reflections) – all while forcing the United States to retain legacy equipment for interoperability.

---

[51] For instance, in late 1997, General Klaus Naumann (who chaired NATO's military committee) observed that the U.S. military was becoming high tech with such "unparalleled velocity [that] one day we will see a disconnect between US and European allies." See James Asker, ed., "Washington Outlook," *Aviation Week and Space Technology*, October 6, 1997, p. 23. But see Gordon Adams, Guy Ben-Ari, John Logsdon, and Ray Williamson, "Bridging the Gap: European C4ISR Capabilities and Transatlantic Interoperability," CTNSP Defense and Technology Paper No. 5, October 2004, Washington D.C (National Defense University).

There is clear advantage even for DoD in extending its cyberspace over NATO allies as they catch up and build their own enterprise infrastructure.[52] Since they are trusted partners, their access to DoD's information stores and processes is likely to be smoother and better exploited than what DoD can expect from friends in the third world.

But how?

It would be easiest on the United States if it built the core elements of one enterprise architecture and had the Europeans plug their equipment into it. For instance, the United States could illuminate the battlefield for everyone everywhere to more or less the same degree, and lend its friends assets (for example, small unmanned aerial vehicles) for their use in critical spots, such as in the neighborhood of their own forces' location. The marginal cost to the United States of extending cyberspace would be modest. But will the degree of dependency implied by this arrangement be acceptable to partners who still enjoy a high degree of military autonomy and a Cold War legacy of rough parity with U.S. forces (especially in ground combat)?

A softer approach is to specify the minimal systems requirements and technical architecture in sufficient detail to enable plug-and-play. Then NATO allies would supply the basic components that would permit them to light their own way when and where they needed it. This would foster interoperability and reduce, albeit not entirely, the systems integration costs that the NATO allies might bear. It would lend allies more scope for their own systems development. Yet this arrangement is still asymmetric, not to mention technically challenging (hence "plug-and-pray"). Architecture is a poor substitute for equipment. Exactly how systems integration requirements will constrain and channel the specifics of the networks, platforms, and weapons away from what NATO allies would otherwise buy is a big unknown – but they are likely to have such effects anyway.

A third choice is to wait until NATO allies reach a high enough level up so that enterprise infrastructures on both sides of the Atlantic can

---

[52] For a fuller treatment, see David Gompert, Richard Kugler, and Martin Libicki, *Mind the Gap: Promoting a Transatlantic Revolution in Military Affairs*, Washington (NDU Press), 1999, chapter 4.

be stitched together as peers of one another. Such an approach, while politically more correct, may nevertheless generate a daunting integration problem. If military systems have not been built with sufficient flexibility, integration may be expensive, incomplete, and failure-prone. If technology keeps advancing, NATO allies may *never* come close enough to the United States to make peer-to-peer integration possible.

The most politically correct approach would have the militaries on both sides of the Atlantic build a NATO-wide enterprise architecture together. Sitting and waiting for a complex multinational effort to reach fruition, though, may postpone the integration of DoD's own infrastructure longer than anyone in the United States wants. And while it is still not entirely inconceivable that America and Europe will someday fight shoulder to shoulder in Europe, Africa, and the Middle East, as they already do in Afghanistan, wounds from frictions over the Iraq war may persist. Anyway, operations in the Americas, the rest of Asia, or at home lack similar prospects for coalition action.

If this problem is ever solved, the solution may require some mix of all four approaches – but, clearly, how a NATO-wide architecture is to be built is no trivial matter to the political cohesion of the alliance. As such, the broader issue of NATO integration suggests that conquest in cyberspace, rather than an end in itself, may be just the newest requirement to keep an existing alliance from fraying unnecessarily.

6.6 Conclusions

This chapter described a logic and some mechanisms by which owners of dominant information infrastructures can leverage them to create coalitions from which they derive disproportional or at least valuable benefits. To be fair, at this point, actual examples of such melded infrastructures are difficult to find, but today's coalitions offer precedents for all the basic elements. As cyberspace continues to grow more consequential, the potential for such interactions can only increase.

Asymmetric coalitions in cyberspace are unlikely to emerge in the near future, and perhaps may never emerge. Coalitions that seemed so attractive or at least necessary at their outset may evolve to become a burden on their owners; much of the crisis that the Japanese economy has undergone after its 1980s bubble burst stemmed from the difficulties in unraveling what were tight webs of mutual dependence among *keiretsu* members. Nor are the dynamics of cyberspace necessarily so obviously attractive to all parties. The owner of the dominant infrastructure may be reluctant to give away information, information services, or connections that were acquired at such expense, even if such generosity in no way diminished the owners' assets. Recipients may be especially wary of dependence and go to great and ostensibly illogical lengths (such as Europe's Galileo satellite navigation constellation) to duplicate a capability that they could have gotten for free (via GPS). Or they may grab the offerings but take care not to rest any important capabilities on the premise of continued enjoyment. Alternatively, independence may last until the first budget crisis – after which point dependence becomes just one of many risks of getting without paying.

The binding mechanisms are both overt and covert. The overt ones arise from the actual information, information services, and connections that the dominant partner brings. They bring out the real usefulness of the relationship for its members – one they partake of or disdain with their eyes open. The thousands of adjustments that must be made by participating infrastructures are the covert binding mechanisms; at some point in the process, one's own infrastructure has been subtly loosed from one's grasp – it may even have unprogrammed behaviors arising from such changes. Restoring its earlier, more specific functionality may become daunting. The most potentially thorough, but at this point least understood, binding mechanisms arise at the semantic level as the categories used within the dominating infrastructure to process information bleed over to all the other infrastructures.

Might matters reach the point where no partner can think about anything complicated without thinking as its partners would? Even the mere act of collecting information is influenced by the words invented to categorize and thus hold it. What it takes to reach such a point is an exercise left for our children.

The result of friendly conquest may not necessarily be power as currently recognized, that is, the ability to make people do what they would prefer not to do. Yet, it may result in unwarranted influence over what they freely decide to do – unwarranted in that friendly conquest's influence exceeds what objective factors of cost and benefit would have dictated.

7

Friendly Conquest Using Global Systems

Two thought experiments may help flesh out the idea that an enterprise – in these cases, a nation – can exercise great influence by creating an enterprise cyberspace that persuades others to link up, join in, and follow along. One is built around a U.S. database of geospatial data. The other is a global identification system, putatively motivated by the desire to detect terrorists.

In presenting these cases, we make no claim that U.S. influence on the world *in these realms* could or should be exercised solely through cyberspace. The general edge that U.S. companies have in the satellite construction, software, and even biometrics would lend them influence without it. The size of the U.S. economy and the size of its national security investments alone would lend it considerable influence in setting the patterns for such systems.

Nor is either prospect necessarily likely. Concepts of information sharing – that lately discovered virtue – that underlie the discussion about geospatial data run antithetical to the everyday norms of the intelligence community, which owns the data. Similarly, despite polls[1] taken in the wake of September 11, 2001, that showed substantial majorities of the

---

[1] A CNN/*Time* poll taken in the weeks after September 11, 2001, found majorities favoring warrantless telephone wire-tapping, unlimited jail time for "terrorists" without trial, interception of every e-mail message, and a national identification card (57 percent favored having everyone carry one); see Michael Elliott, "A Clear and Present Danger," *Time*, October 8, 2001, p. 29. A contemporaneous Harris poll showed that 68 percent of adults surveyed (by telephone) favored the adoption of a national ID card for all U.S. citizens. "National IDs Won't Work," *Business Week*, November 5, 2001, p. 86.

public in favor of a national ID system, opposition to such a notion is visceral and the Bush administration has gone out of its way to deny having any such thoughts.[2]

Nevertheless, what follows might illustrate how influence over others can be pursued through cyberspace.

7.1 Geospatial Data

One reason that a global utility for geospatial information might profit everyone is that geospatial information – natural landforms, human artifacts, political declarations – is general information. General information is the same for everyone – the height of Mt. Everest is constant and public – and is of potential interest to all. By contrast, most health data, epidemics aside, are about an individual and of little proper interest to others. The more general the information, the greater the economies of having it centrally acquired, organized, and disseminated. The more specific the information, the fewer the benefits of centralized management, and, in some cases (as with health information), the greater the problems (such as lost privacy) of doing so.

The logic of a unified global geospatial database, however, confronts the reality that sovereign nations are defined geospatially, and in no other way, even today. To map something is, in a deep sense, to own it.[3] The exclusive right to map one's own terrain feels like a matter of national sovereignty as well. Soviet mapmakers felt they had the right and, indeed, the duty to portray sensitive sites as being several miles away from where they actually were. This perception of national proprietary interest is in no way dulled by the possibility that those who would map another's territory often do so for less than altruistic reasons.[4] The same holds for imagery – from which most modern mapmaking starts. Taking pictures of another's territory was, for decades, considered an act of espionage,

---

[2] For instance, Richard Clarke, who headed White House efforts to defend the national information infrastructure in the early George W. Bush administration, noted that "he could not name one official who supports the idea [of a national ID card] as proposed [by Larry Ellison, CEO of Oracle]." Robert Lemos, "Bush Adviser: Terror a Real Threat to Tech," http://news.com.com/2100-1001-829485.html, November 8, 2001.

[3] See, for instance, Stephen S. Hall, *Mapping the Millenium*, New York (Random House), 1992.

[4] As portrayed, for instance, in the movie *The English Patient*.

almost tantamount to war – and so such instruments are referred to as *spy* satellites.[5]

The United States government, not surprisingly, possesses a great interest in geospatial information. Its National Oceanographic and Atmospheric Administration populates the world's grid with constantly changing temperature and humidity information. The U.S. Geophysical Service (USGS), together with the U.S. Census Bureau, maps this country. The U.S. National Intelligence and Geospatial Agency (NGA) maps everyone else's. All geospatial information can be digitized as pure data and thus stored and transmitted as pure data. Maps are coded as vectors (such as for roads and topographic features) and georeferenced databases; imagery is coded as pixels. NGA, with its imagery collections, may well own more bits than any other organization in the world – and the overwhelming bulk of it, especially imagery, is for certain eyes only. There *are* national security advantages in others not knowing exactly how good one's collectors are, how accurately the world of interest has been photographed, orthorectified, mensurated, and assessed, and what parts of the world (essentially, which facilities) are subjects of especial interest to U.S. spymasters.

Geospatial information can be considered a treasure trove – but it can also be considered a potential information service of use to others and, as such, a foundation for a specific type of cyberspace. It could be the system that everyone else with interest in such things wants to spend time in: read from, write to, invest in, and thus adapt to. The U.S. geospatial database could well be that cyberspace, were enough of it made accessible, something that would require (1) complete or near-complete digitization of information,[6] (2) a formatting convention that could be widely read, (3) enough security to prevent others from corrupting files or attacking the links to such a cyberspace, and (4) potentially universal access with adequate reliability and bandwidth.

Premium access to such a database may be viewed as a gigantic bargaining chip that would permit the United States to get data held by

---

[5] See Mark S. Monmonier, *How to Lie with Maps,* Chicago (University of Chicago Press), 1991.

[6] Actually, both USGS and NGA have lagged in digitization, due in part to the enormity of the legacy files they have to convert and the slow arrival of software capable of bringing computer-generated maps up to aesthetic standards that the professionals in these agencies consider proper.

others. Notionally, everyone would be better off by such a trade.[7] After all, obtaining already-collected data from others may cost less than reacquiring it oneself. Cyberspace, for its part, makes it easy to carry out both the exchange and amalgamation of such data.

Cyberspace could make such a database, and the standards by which it is organized quite influential. The ease with which others can get information may, for instance, shape their own collection around what was lacking or needed validation in the database. Storing information in cyberspace lets other users annotate the data they see (such as adding historical or other details tagged to a location or facility) and automatically return such annotation to a common pot. So, annotations can themselves be annotated. Maps make very good whiteboards for collaboration, as eons of generals have discovered. Others could also offer their own data with the hope that third parties, even elsewhere in the U.S. government, could supply annotations in turn.

This model would seem to require some faith in the kindness of strangers. Why not take, take, take, and give only that which is absolutely essential? Some may well try, but raw selfishness is not an obvious outcome in all cases. There are communities bound by the soft reciprocal obligations of gift giving and its attendant prestige. Science provides one example. The Linux open-source community provides another. To those who cite the hardness of heads typically found in national security, and especially the intelligence communities, one may counter that many nations do stand up to be counted when it comes time to contribute to multinational coalitions, such as in Kosovo and Afghanistan. True, U.S. leadership was the sine qua non of coalition formation in those two countries, but that would also be true in the arrangement discussed here.

The formation of such a community effort would give the United States government worthy benefits. It would at least be able to exploit the data

---

[7] In practice, the United States is better off only if the value of the information it gets back (compared to the data it already owns) exceeds the cost of ascertaining its quality and reliability. In addition, the marginal value of adding any one partner may be vitiated if the partner becomes a conduit for data to free riders. Whether or not other countries find such a deal worthwhile may also depend on whether the United States collects all the data of interest to them. If not, and such countries feel they have to survey their lands anyway, the savings to them from linking up may be modest.

of contributors for having gotten it – and in a format that is likely to be consistent with what it already has. It also profits from having others collect and format information based on or at least consistent with the kind of inquiries it deems important. If customers find this geospatial commons of value, they are more likely to regard such a commons protectively and, with it, maybe even extend their solicitousness to the owner's prerogatives (if not in the data than at least in the procedures that collect and manipulate the data). U.S. government geographers may be able to influence the evolution of geospatial knowledge bases and their technologies. Their ability to do so would be based both on their deep knowledge of the global system (notably its rationales) and the likelihood that the innovations they propose are more likely to be implemented than those that come from the outside. Even digital mapmakers can highlight what they want others to see and divert attention from what they would rather keep hidden[8] – not by erasure or shading but by using a data model that makes certain data easier to find than others, and more intuitive to put together. For instance, a postwar database of an area can have more detailed and geospatially precise[9] findings of weapons of mass destruction, which are then indicated more precisely and in better-linked ways than deaths due to collateral damage in warfare. Choices over what is grouped together can conflate major and minor elements; for example, calling both chlorine containers and nuclear materials weapons of mass destruction exaggerates the former or demotes the latter.[10]

---

[8] Mark Monmonier has detailed the controversies associated with a global projection system that made imperial Europe look much larger than colonized Africa. Mark Monmonier, *Rhumb Lines and Map Wars: A Social History of the Mercator Projection*, Chicago (University of Chicago Press), 2004.

[9] It is often difficult to map something if the error in estimating its location exceeds the effective accuracy of the map as a whole.

[10] Nevertheless, the cognitive influence that can be transmitted via geospatial database has limits. First, many geospatial terms have long histories and cannot be willy-nilly redefined or promoted. Second, modern geospatial databases are built from layers, which allows users to generate maps from the layers that *they* feel are important (by contrast, a map shows information the mapmaker deems important). Third, most of what the United States can contribute to a global geospatial database will come from what it can gather from space. Distinctions that cannot be made from space (for example, is this a hospital, school, or office building?) require local knowledge, almost all of which will be supplied by whoever governs the area in question.

Can the United States dictate standards[11] for such a database? Should it? As a practical matter, the U.S. government has to keep an open mind on how others want to format geospatial data; only some of the decisions made on data standards carry broader architectural and cognitive ramifications. Given the dominance of private firms in selling geospatial software, even the United States needs to respond to the market-led evolution in data formatting specifications, which are increasingly expressed via XML tag sets or toolkit algorithms (via APIs). Those who own the cyberspace should be willing to make the requisite investments, and be lucky enough to ride or override technological or political trends that would usurp its status. It also helps to have a fine sense of how far those owners can exercise discretion (for example, by not hosting certain data elements within a model or by excluding certain contributors) without inducing others to form countercoalitions. The desire by others to see the adoption of neutral and public standards for the description of data is strongly felt.

Yet, there may be ways to follow and even advocate open standards without losing control over how information is perceived. One way is to develop and promulgate compound concepts (for example, for "ground cover") whose data structure is consistent with one's own perspective on how to view the world. Another is to develop applications that can be accessed by others, but are housed and controlled on the leader's site. These applications could read and write standard terms, but they (1) require privileged access to the site; (2) may, as a quid pro quo, monitor how such applications are used (for example, by whom, when, and to solve what problems, and applied to what data set); and (3) manipulate data structures in particular ways. Applications that permit one party to talk to another through the leader's facility (such as fly-throughs over terrain) may be particularly attractive and, for that reason, likely to bind users. Those who would offer such applications, however, have to be agile to stay ahead of corresponding free Web offerings such as Google

---

[11] Standards for describing geospatial features have been in use for decades both on the military side (for example, MIL-STD-2525B) and on the civilian side (such as the National Institute for Science and Technology's [NIST] Spatial Data Transfer Standard); see Martin Libicki, *Information Technology Standards: Quest for the Common Byte*, Boston [Digital Press], 1995, pp. 322–4.

Earth or Microsoft's Virtual Earth. What is value-added today may not be value-added tomorrow.

7.1.1 Coping with Commercial Satellites

Modern GIS systems integrate semantic map data and imagery, and the one that I have just sketched would be no exception. Although imagery has virtually no semantic content apart from pixels, a global utility that hosts real-time imagery is one on which the United States can exert influence in pursuit of its values. Having such a capability provides ways for the U.S. national security community to cope with its concerns over the proliferation of precise remote sensing satellites and even exploit their presence.

Until 1999, only one nonmilitary satellite, France's Spot, could capture images that could resolve details down to 10 meters. Had Iraq gotten such imagery at the wrong time, it may well have seen the Left Hook coming at them prior to the ground campaign of Desert Storm.[12] The launch of Ikonos in September 1999, however, opened up an era of commercial spy satellites that could resolve details down to .82 meters – good enough to report on the condition of the EP-3 aircraft grounded in Hainan, China, in 2001. Next, DigitalGlobe, essentially an Israeli company, launched Eros with similar resolution. In late 2001, a third satellite, Quickbird, improved the resolution standard to .61 meters. Future high-end models may well reach down to the quarter-meter mark.[13] Surrey Satellite Technology sells satellites that can achieve 6 meter resolutions but for only $20 million, launch costs included. Their RapidEye contract to supply five such satellites for German use runs $35 million, launch costs and optics *excluded*.[14] India launched a surveillance satellite with

---

[12] See Vipin Gupta and Lt. Col. George Harris, "Detecting Massed Troops with French SPOT Satellites: A Feasibility Study for Cooperative Monitoring," presented to the "Secret No More: The Security Implications of Global Transparency" conference, Washington, May 21–2, 1998.

[13] Space Imaging has acquired the license to fly a satellite with .4 meter resolution, and has applied for a license to fly one with .25 meter resolution. See www.spaceimaging.com/newsroom/2003_new_policy.htm.

[14] Frank Morring, Jr., "In-Orbit," *Aviation Week and Space Technology*, June 21, 2004, p. 27.

2.5 meter panchromatic (black-and-white) resolution for a total pro-gram cost, including other expenses, of roughly $50 million.[15] Although synthetic-aperture radar satellites that see at night and through clouds are more expensive, they, too, are becoming faster and cheaper. Germany's TerraSAR-X, scheduled for launch in late 2006, boasts 1 meter resolution and should cost $150 million.[16] If such price trends come to character-ize the commercial market, a future in which tens or even hundreds of satellites ply low-earth orbits cannot be ruled out.

A narrowing gap between what billion dollar investments by U.S. three-letter agencies buy and the pictures that can be purchased within many people's credit card limits worries some in the U.S. national secu-rity community. True, DoD satellites, cumulatively, still have a plethora of advantages: greater acuity, multi- and hyperspectral capabilities, and radar imaging; rapid transmission of imagery from orbit to analysts and increasingly directed to warfighters; and a large and experienced cadre of photo-interpreters. But even if a gap survives, the importance of the gap may not. Take a 20:1 gap. The person with a 40 meter satellite may not be able to make out anything of military relevance, while the person with a 2 meter satellite can see military formations. If both advanced twenty-fold, the commercial user has the 2 meter resolution ("I see something large enough to be a tank"), while the government user would enjoy .1 meter resolution and yet have an edge only in identifying the specifics ("I know which division the tank belongs to"). The fact of the formation – its location and composition – is no longer a secret.

The U.S. government has responded to the shrinking gap in several ways, besides by improving its own capabilities. One is to exercise shutter control over those satellites built in whole or in part with U.S. com-ponents – a strategy that, at best, applies to only some satellites and causes customers to perceive U.S. suppliers as less reliable. Another is to do nothing until there is a conflict, and then block-purchase all pho-tographs taken over the battle zone – a strategy used to keep Ikonos

---

[15] See www.isro.org/Cartosat/page3.htm for a description and http://www.globalsecurity. org/space/library/news/2005/space-050502-irna01.htm for program costs.

[16] See www.terrasar.de/en/imp/hist/index.php.

imagery of Afghanistan off the market.[17] A third is to declare, at least sotto voce, that satellites that supply adversaries with battlefield imagery are, as such, hostile and may be disabled either temporarily (by jamming uplinks to the satellite or computer network operations, for example[18]) or permanently. In all three cases, the chance to exercise control requires knowledge of what disquieting imagery is or at least of what such imagery could be taken. Otherwise, assurance can come only by shutting down *all* suspicious imaging satellites for the duration. This may not sit well with nonaligned countries that possess their own surveillance satellites or that feel protective of their satellite-owning corporations.

Over and above such controls on proscribed behavior, economic carrots have also been employed to keep the industry on friendly and persuadable terms. The NGA, in 1999, made much of its promise to buy up to a billion dollars worth of imagery over the next five years; the promise helped at least one satellite company clinch financing.[19] But, although military and intelligence purchases worldwide account for as much as 40 percent of all imagery sales, actual purchases by NGA fell far short of promises.[20]

A global commercial remote sensing industry truly independent of the U.S. government and universally perceived as such may not necessarily be a bad thing, though. A business model for the sector could spur innovation in the use of space-based imagery and processing tools. Faster innovation may in turn spin off technologies and techniques that are

---

[17] See *Space News*, October 22, 2001, p. 6.

[18] Hackers could, in theory, interpose harmful or at least useless commands from the ground station to the satellite as well as introduce erroneous artifacts or erase valid ones in image between capture and delivery. If not that, their efforts may still reveal that battlefield imagery is being (1) acquired and (2) passed to adversaries directly or indirectly. In truth, success is hard to imagine against any opponent but a naive one. Space operators are already heavy users of air-gapping and encryption, and those who have not yet done so could easily add digital signatures to this repertoire.

[19] At the time of the offer, a representative from ITT said that the NGA's promise was the key to its investing in EarthWatch. See *Space News*, April 26, 1999, p. 3.

[20] This program has had a difficult time getting started. Annual spending ran closer to $40 million a year, with an interim goal of $80 million a year, according to the director of the National Reconnaissance Organization (NRO); see Robert Wall, "NRO Wants License to Take Risks," *Aviation Week*, September 11, 2000, p. 38.

more likely to redound to the U.S. defense sector than to anyone else's. True independence may persuade others that the images are credible and do not show up only when convenient to U.S. foreign policy interests. Insofar as transparency favors the U.S. position in world affairs (that is, that we are the good guys with nothing to hide), credibility is good.

7.1.2 Manipulation through Cyberspace

Building a living geospatial repository in cyberspace could add new ways to approach the "problem" of burgeoning remote sensing capabilities.

The government's pricing of Landsat 7 satellite imagery provided an inadvertent example. In authorizing the satellite, Congress mandated that its imagery not be sold at a price that exceeded the marginal cost of reproducing it, or roughly $600.[21] Although scientists applauded,[22] the remote sensing industry argued that the dictum threw a wet blanket on their sales prospects. However, their griping should be understood in context. Landsat's 15 meter imagery is hardly in the same class as the .6 to .8 meter imagery of commercial satellites. Low-cost "before" imagery of an area may also boost the market for "after" imagery of the same place. Were prices to reflect what it normally costs over the Web to store and transmit the 280 megabytes that a typical Landsat image contains, everyone would want some Landsat pictures of his own, creating a great "before" market that forces commercial satellite operators to ask themselves exactly where they are adding value. Such a market is possible only through cyberspace.

Establishing a site in cyberspace for imagery may well influence the *internal* structure of the industry as well. Will it be satellite operators or

---

[21] As anyone who has looked at Microsoft's Terraserver 2 meter imagery can attest, raw imagery is not always usable. Thus, the $600 price may include the cost of cleaning up images by hand. Otherwise, such a high price can be neither justified by storage (the imagery has to be stored on network-linked servers to begin with) or transmission (even the 280 megabyte files that characterize Landsat shots can be transferred through a $15 per month DSL account modem in under an hour – and for pennies).

[22] A consortium of Ohio universities has announced plans to purchase Landsat 7 imagery and make it network-available. See *Space News*, May 17, 1999, p. 16. See also Beth Lachman, *Public-Private Partnerships for Data Sharing: A Dynamic Environment*, RAND draft, DRU-2259-NASA/OSTP, April 2000.

postprocessors, for instance, who have the market power? In one hypothetical case, customers first order pictures from a satellite, and only then have them postprocessed by any of a number of service providers. Here, an overt link between customer and satellite owner makes it easier for the United States to identify and punish those who may take pictures of proscribed areas.[23] Alternatively, if postprocessors are in charge, customers would give *them* imagery orders – or to further the case, a more general request to elucidate a phenomena or track an event – and they would, in turn, find the most convenient satellite to take the snapshot. This would insulate any one satellite owner from reprisals; enforcers would have to acquire the rogue pictures surreptitiously and determine which satellite with what technical characteristics was overhead when the picture was taken. If it is the U.S. interest that the satellite owners rather than processors influence that pictures are taken – and neither market model has yet to be locked in – then the adroit insertion of U.S. capabilities (including photo interpretation, archives, and analytic services) into the market may tilt the market accordingly. Making such services available through cyberspace seems the easiest way to do this.

U.S. influence over a cyberspace built around imagery may also persuade non-U.S. vendors to play by U.S. rules and thereby offset temptations to serve potential adversaries. Perhaps this is more easily done than trying to tilt the market in favor of U.S. vendors. One inducement might be that compliant satellite owners[24] could avail themselves of images and services from the U.S. government; cyberspace makes this possible at nearly no cost or lag time. Software could also be made available to satellite owners that would help them integrate their own images with archives, in order to support before-and-after analysis. Other value-added services might include feature recognition, mensuration, display, and integration

---

[23] One way to detect whose pictures are being passed around is to mandate that the satellite owners insert a chip that would watermark the picture (by manipulating low-order bits in the image). Analysis of the image (actually the file from which the image is generated) would indicate which satellite took the picture. The art of making a watermark unobtrusive and impossible to eradicate (even by the satellite owner) is not simple, though.

[24] Information services can be limited to those with images that provably come from compliant vendors. However, if access to images is used as the carrot, how could one tell that customers of compliant vendors are also not customers of noncompliant ones?

with other databases (of port terminals, for example). Perhaps members that register their own imagery could then sign up for a constant stream of value-added information as it becomes available (for example, from subsequent U.S. images or other U.S.-generated databases).

Customers would add value to their own images by exploiting services hosted on this cyberspace – but if the possession of such a cyberspace is to be influential, it will be necessary to require imagery to be sent to the value-added service and not the other way around.[25] The degree to which having such a privilege motivates compliance among satellite owners would probably be proportional to the *value* of such a privilege vis-à-vis commercial offerings.

7.1.3 Getting Others to Play the Game

The current policy of maintaining control by limiting the number of noncompliant satellite owners assumes a world of few vendors with good satellites (such as those with submeter resolution). If the planet is girded by hundreds of satellites with resolutions in the 1–5 meter range, chances are that noncompliance would be a core element of *some* satellite owner's business model. Keeping sensitive areas hidden, at least without resorting to coercion, would be nearly impossible. The next best approach is to turn the balance of transparency to U.S. interest. Such a strategy builds on the premise that while one satellite may reveal nothing that a defense satellite cannot reveal faster and better, many working together could keep any, even if not every, spot under continual if not continuous surveillance. With so many satellites, multiple images may be taken well in advance of specific imagery needs (for example, many people can take any one shot, and the value comes from being able to build an interesting program of shots). So how can the United States persuade satellite owners to do

---

[25] Under UN convention, a country whose territory is commercially imaged has the right to buy the image at the same price as the original customer did. This, it seems, would forbid such images being taken in secret. Thus the fear that U.S. intelligence would learn what commercial imagery is of interest to others is beside the point. Is there much intelligence value in knowing that a nation is buying images of its own territory? Perhaps, if the country is trying to figure out whether other satellites can see what it may be trying to hide.

its scanning for it – either to exploit results of what they find (such as something that has been tracked or a rumor that has been disproved) or to cue the United States to turn on its national assets to better identify and characterize the finding?

Consider a collective effort to track ships. Most ships broadcast their position periodically and have transponders to answer queries automatically. The ones that do not are interesting. Can satellites provide a way to police the seas against rogue ships, pirates, and polluters? The effort would require data not only from the coordination of imagery collection but from other databases (such as those of ship types and port characteristics), commercial registries, shore-based sensors, ship logs (such as those that note unexpected sightings), and, one day, UAVs on border patrol. On the high seas, though, satellites come into their own as trackers. There, anything imaged that is not water or ice is automatically noteworthy.

Cyberspace could be the medium through which the United States offers the foundation (including coordination tools, data exchange forums, and such) of a ship-tracking service in the hopes that satellites will contribute to it when over the oceans.[26] Presumably, the United States has an interest in rapid public posting of key results, perhaps to excite customer interest. It will want to encourage each satellite to look where others have not looked rather than have each look for themselves where others sense something. But if it wants moving targets (such as ships) tracked, it should also want various satellite operators to be interested in posting information. Other owners would then to be able to confirm and update tracks, in turn, passing their updates along.

In this scenario, the U.S. government would be a broker – identifying and providing the starting location for objects of interest, acquiring and storing the information collected on it, amalgamating ancillary sources of information, deriving probability estimates of next sightings, extracting key features of interest and adding value to its collections, and providing a mechanism to prompt others to contribute. Capacious storage, robust networking, sophisticated tools, stronger security, and better archives make for a good broker. No entity is better placed than the U.S.

---

[26] Satellites store up power when they are not taking pictures. Thus taking a picture over the oceans would not be entirely free even if the satellite is otherwise idle.

government to assume this role. Having other commercial satellite owners contribute to and draw from the cyberspace is a position of considerable influence.

7.1.4 Some Conclusions about Geospatial Services

Creating a global utility of geospatial information services provides a vantage point from which the United States, as owner, can both create and shape a community. Shaping can occur directly, as when access is contingent on friendly behavior. It can occur indirectly, as when interactions are the vehicle by which everyone's understanding of geospatial data acquires a U.S. coloring. In the case of commercial remote sensing, such services can help the industry evolve in ways aligned with U.S. interests in global transparency and law. Realizing such potential, however, requires recognizing the power of generosity, contrary to current practices,[27] and recognizing that information can be more valuable when given away than when hoarded.[28]

7.2 National Identity Systems

What avenues exist by which the United States can influence the global evolution of identity systems (assuming, of course, that it wants to support a system by which unique known identities can be reliably associated with individuals)? This section approaches the question with some rationales for a national identity system[29] and a brief sketch of *one* such system to satisfy such values. This then sets up the heart of the section: how such

---

[27] After September 11, 2001, NGA, for instance, decided to retard the release of global information on land elevations as collected by the space shuttled radar topography mission (SRTM), especially outside the United States. See *Space News*, February 4, 2002, p. 11. The first unclassified material was released in early 2005. Frank Morring, Jr., "In Orbit," *Aviation Week and Space Technology*, February 28, 2005, p. 17.

[28] To some extent, such thinking has already started to appear in the commercial world. Google Earth (earth.google.com) is a downloadable application that provides users space-based imagery at various degrees of resolution (it has a correlated mapping service). One way to make money from such a service is to charge for listings (the pizza restaurants within a mile of Grand Central station, for example). Microsoft Networks (MSN) has a competing service, Virtual Earth.

[29] "National," here, includes international travel documents recognized (or issued) by nations.

values may be globally transmitted through cyberspace. To repeat, the purpose of this section is not to argue for or against such a system or its inevitability, but to explore how values embedded in one nation's system may influence values embedded in the systems of other nations in the course of assuring interoperability among them.

7.2.1 Two Rationales for a National Identity System

Three aspects of the September 11 attacks suggest that prosecution or preemption based on intensive surveillance would have been of limited or no help. Prosecution, at least of the hijackers, was preempted by their suicide deaths. Only two of the hijackers were on any watch list; there was no direct[30] evidence available to the U.S. government that would have merited surveillance of the other seventeen.[31] Conversely, none of the nineteen were U.S. persons (that is, citizens or permanent residents). A legal regime that, broadly speaking, differentiates between those who are and those who are not U.S. persons can make politically attractive trade-offs between security and privacy and may be effective – as long as this distinction characterizes future attacks.[32]

Identity systems essentially do two things.

First, they permit the reliable association of individuals with those aspects of their history that indicate that the government can treat them differently. Some individuals, for instance, have arrest records and are recorded as such by fingerprints in FBI-maintained files. Others are aliens

---

[30] SRD, a data services corporation, argues that most of the nineteen could have been linked to the original two if there had been an intensive data tracing regime in place prior the hijacking. Such a regime could have unearthed transactions associated with the two hijackers already known to be potential terrorists, and by so doing (through multiple iterations) could have eventually revealed names of all nineteen. See Markle Foundation Task Force, "Protecting America's Freedom in the Information Age," http://www.markle.org/downloadable_assets/nstf_full.pdf, October 2002, p. 28.

[31] See *The 9/11 Commission Report: Final Report of the National Commission on Terrorist Attacks upon the United States*, New York (W. W. Norton), 2004; for a rejoinder, see Richard Posner, "The 9/11 Report: a Dissent," *New York Times Book Review*, August 29, 2004, p. 1, ff; idem, *Preventing Surprise Attacks: Intelligence Reform in the Wake of 9/11*, Lanham (Rowman and Littlefield), 2005.

[32] This is a big *if*. The July 7 and 21, 2005, attacks on the London subway system were apparently carried out by British subjects, not foreigners. Reports immediately following their identification of such individuals stressed how seemingly ordinary their lives appeared to be up to that point.

of one status or another. For instance, the combination of the U.S. visas and the USVISIT program at the border links each visitor (from countries whose nationals need visas to enter the United States) to a pair of fingerprints taken when the visitor applied for the visa. These fingerprints are then taken again and verified as belonging to the correct individual when these people enter the United States. Domestically, a national identity system makes it hard for a fraudster to assume the identity of someone who is registered with his or her biometrics (for example, a facial photograph).

Second, it permits governments to build up a history on individuals. In retrospect, the nineteen hijackers evidenced patterns of activity that may have raised suspicions were they correlated in advance of the hijacking: variant visa status, attendance in flight schools uncorrelated with prior work experience, substantial cash transfers, inquiries into crop-dusters, and buying first class airline tickets. A surveillance system could combine a reliable identification system with a program of recording selected activities. Such data could be analyzed by people working with intelligent software to identify individuals who may merit closer scrutiny. A national identity system would thereby associate individuals with specific identifying characteristics with specific chains of possibly suspicious events.[33]

7.2.2 Potential Parameters for a Notional System

The essence of any identity system is the ability to associate a person with an identity. Identity cards do this by associating the name on (or in) the

---

[33] In the year after 9/11, the Defense Advanced Research Projects Agency (DARPA) initiated a technology development project, Total Information Awareness, which, if successful, would have led to the gathering of an enormous database of transactions carried out in the United States (and, in some cases, overseas). Software applied to this database would infer the presence of potential terrorists. Congress killed the program largely on privacy grounds. Privacy was not its only problematic feature: there is no reliable validating of identity in most commercial transactions, no mechanism to compel the production of everyday transactions data from the businesses that collect it, no algorithms for detecting terrorists with acceptable rates of false positives and false negatives, and no doctrine of what to do when someone suspicious has been identified. The last two criticisms may be made of the checkpoint system to be outlined here – which is why this chapter is not an argument for the system as such but a discussion of how such systems can exercise global influence on behalf of their owners.

card with a set of biometrics (such as a facial photograph) and associating a set of biometrics (for example, what the person's face looks like) with the person presenting the card. The biometrics need not be on the card if they can be summoned from across a reliable ubiquitous network. Thus, the card itself would be no more than an electronic token that links some identifier to two databases. One database would contain biometrics information and some personal parameters (such as birth date, sex, and citizenship status) and would be consulted only to see whether the person has presented the correct identifier. The other database is a record of checkpoint events sorted by identifier.[34]

Biometrics, a person's unique identifying physical features, are only used to correlate identities and individuals consistently (if not necessarily accurately). They include fingerprints, but also iris scans, palmprints, facial features, voiceprints, signatures, and DNA. Collecting them as part of an identity registration process serves two purposes. First, it prevents two people from assuming the same identity by ensuring that if one person is registered with an identifier, a second person, who will invariably have different biometrics, will be carrying a card (or a link to a data record) whose biometric data do not match the first person's. This is known in the trade as one-to-one matching. Second, it also prevents one person from assuming two identities within one database. When the person tries to register a second time under a different identifier, the biometrics[35] he or she presents are compared to the total database of biometrics; if the biometrics match those of someone already registered, chances are the same person is attempting to register a different name. Fraud can rightfully be suspected. This is known as one-to-N matching.

Both tests require careful documentation and authentication of how biometrics were collected, validated, and assigned to a cardholder. The biometrics database would contain a digitally signed record of such authentication. If the card were used as the validation device, then it

---

[34] Existing checkpoints include any occasion (such as an arraignment) where a passport is presented or fingerprints are taken. Potential checkpoints may include airplane and ship travel; vehicle and machinery rental; the purchase of certain chemicals; entry into secure, safety-related areas or government-sensitive facilities; and large cash transfers.

[35] Preferably one wants a biometric, such as fingerprints, DNA, and (probably) iris scans, that is (1) hard to fake or disguise, (2) stable over many years, and (3) sufficiently unique that the odds that a given person's biometric is shared by someone in a very large database (with tens or hundreds of millions) are small enough to be manageable.

would make sense to have the relationship of the photograph and identifier digitally verified.[36]

Such a system can ensure a one-to-one correspondence between registration records and registrants, but it does not suffice for ascertaining that everyone enrolled as a U.S. person is, in fact, a U.S. person. Doing that requires establishing some correspondence between a person and historical documents such as birth certificates or school records, perhaps supplemented by personal testimonials ("I've known this guy for years"). Finally, assigning a U.S.-issued identifier to individuals from countries with indifferent record keeping may not say much about who they were or what they did before coming to the United States or while outside the country.

Linking an identifier to an event might work like this. The card is presented to a card-reading device (CRD) so that the identifier can be read (if a cardless system is used, the identifier could be punched into the CRD). The device passes the encrypted identifier to the biometrics database and a photograph comes back; alternatively, the photograph is read directly from the card. The CRD operator sees the image, on the computer screen or the card as the case may be, and compares it to the holder's face. If there is a match, the event is recorded; if not, the cardholder is challenged. For greater accuracy, substitute fingerprint matching for visual face matching.[37]

If it is deemed important to suppress casual requests for an identification card (for example, by police officers or bartenders), the card may be designed not to have an identifier displayed. The identifier itself would be transmitted to the CRD in encrypted form and decrypted only inside the

---

[36] The optimal number of enrollment points is a trade-off between the desire to minimize the opportunity that one might be compromised and the desire not to inconvenience the user too much. The problem is one of authentication, and the possibility that an enrollment clerk may be bribed or otherwise persuaded to register a false transaction (for example, by falsely certifying someone as a U.S. person). The more enrollment points, the higher the odds that there will be a corrupt enrollment clerk. There are numerous ways to battle the possibility of corruption, such as having two officials sign enrollment transactions or having the system run its own checks automatically (as it does when it looks for duplicate fingerprints upon enrollment).

[37] A sufficiently sophisticated system based on biometrics should be able to return a name and/or an identifier without a card or even a number if the person presents a thumbprint.

biometrics and checkpoint databases. Since the purpose is to log passage for later analysis, no identifier need be returned to the CRD operator unless the card was used for access control. A picture on the card would, however, be useful. Such a picture would not say anything about the identity of the presenter (any more than the presenter's face itself would) but it would permit people to distinguish their cards from those of others if they were mixed up.

7.2.3 Constraints from and Influences over Foreign Systems

A national identity system designed to monitor the activities of foreigners cannot help but have international ramifications and this cannot but influence its overseas counterparts if their citizens can come to the United States without a visa, as citizens of Canada and twenty-eight other countries can do.[38] Because citizens of other countries need visas, such visas in turn would be the document of record and, as such, already in conformance with U.S. standards. Countries that wish to remain in the visa-waiver program, however, might well therefore be asked to generate registration procedures and an ID card that would meet U.S.-required standards, as follows:

- If identifying information is to be captured from the passport, it would have to be electronically readable by U.S. machines. Some passports are already machine-readable, but U.S. passports do not yet transfer enough bits to validate a digital signature or photograph in the process. Were a national identity card to contain a radio-frequency identification device (RFID)[39] for the purpose of

---

[38] Citizens from the following countries do not need visas for stays of less than ninety days: Andorra, Australia, Austria, Belgium, Brunei, Denmark, Finland, France, Germany, Iceland, Ireland, Italy, Japan, Liechtenstein, Luxembourg, Monaco, the Netherlands, New Zealand, Norway, Portugal, San Marino, Singapore, Slovenia, Spain, Sweden, Switzerland, the United Kingdom, and Uruguay. Some of them, such as Australia, require visas of U.S. citizens.

[39] As of this writing, the U.S. State Department is contemplating putting an RFID chip in U.S. passports so that they can be read electronically. See Kim Zetter, "Feds Rethinking RFID Passport," *Wired News*, April 26, 2005, available at www.wired.com/news/privacy/0,1848,67333,00.html?tw = wn_story_related. Many have objected to this feature, fearing that third parties (including criminals or terrorists looking for

permitting unmanned or even surreptitious checkpoints, then the option of accepting other countries' travel documents would be limited. All other methods of electronically reading the cards require that the card be presented to a manned CRD. If the information is captured through other means (such as having people give out their name or identifier) and validated over the network, then everyone's networks would have to interoperate for identity to be validated.

- Foreign issuers would have to meet U.S. standards for authentication and for the capture of biometrics information. One visa-waiver program country that allows people to register incomparable biometrics creates a loophole for fraudsters.
- If everyone were limited to a unique identifier, then every visa-waiver country would have to ensure that its registrants were unique both nationally and within the entire database of cooperating countries. Such countries must therefore (1) possess biometrics-linked registration systems that are capable to catching unauthorized duplicates and (2) check applicants against the files of other countries based on such biometrics. Because of the latter requirement, every country would have to collect at least one globally agreed-on set of strong biometrics.[40] Thus there have to be technical standards for encoding biometrics in compatible ways. International networking would also have to be sufficiently robust to permit these checks to be completed while the applicant is being processed; otherwise someone could get two identifiers by showing up in two countries within hours – with neither having any record of the other one's biometrics in time.

Americans in crowds overseas) could read such passports surreptitiously, invading privacy or worse. Bruce Schneier, "Does Big Brother Want to Watch? Passport Radio Chips Send Too Many Signals," *International Herald Tribune*, http://www.iht.com/ articles/2004/10/04/edschneier_ed3_php, October 4, 2004. In response, the State Department said it would impose new security techniques, require encryption for data transfers, and ensure that passports contain a metallic layer. See http://www.epic.org/ privacy/rfid/. The potential to read RFID cards surreptitiously enables the government to establish checkpoints that are mobile (perhaps carried by officials) and quasisecret (that is, they are not obvious when people are passing by them); this is what may make them controversial.

40 Countries can use individual numbering schemes as long as the identifiers they issue indicate some sort of country code. Dual-nationality citizens can be accommodated as long as their various identifiers are linked in databases.

- The United States may want to construct or refine a watch list based on passing a set of biometric indicators (a facial photograph if nothing else) past its entire biometrics database. To ensure that no one could enter the country without being prescrubbed (that is, vetted in advance), it would need to be able to pass this set through a similar file containing the passports or national identity documents of all visa-waiver countries.[41]

These are not trivial considerations. Even something as straightforward as associating an individual with a set of traits is not without controversy. What accommodations, for instance, should be made for people who object to offering a biometric on religious or privacy grounds? Conversely, the United States would not collect certain types of information normally collected elsewhere, such as race and religion.

The requirements for reciprocity will also complicate setting a consensus on design standards, especially if the United States wishes to inhibit casual reading of such cards. If the United States would read the passports issued by other countries, these countries may well insist on being able to read the normally encrypted U.S. cards. Such countries could be given the U.S. decryption key but then they must protect the key or keys as well as they are protected here. Yet the more overseas locations have a decryption key, the greater likelihood the decryption key will leak back to the United States.[42]

Real influence is exercised through cyberspace when data-sharing issues arise. Terrorism and transnational crime are international problems. The September 11 plots were hatched in Germany and supported by travel within Spain and Malaysia. Richard Reid, the would-be shoe-bomber, was a British subject. Zacarias Moussaoui held a French passport. The United States would be helped if noncitizens were monitored in friendly states and the data shared among systems. But then the United

---

[41] A traveler from a country not in the visa-waiver program would need a visa in advance. The information required to generate such visas could also be linked to the national identity registry.

[42] An alternative for countries that wish to inhibit casual perusal of the national identity cards is to issue two flavors: one for domestic travel containing an encrypted identifier and one for overseas travel where the identifier can be read by any suitable card-reading device.

States itself must be prepared to share data and may not be able to exclude U.S. citizens from scrutiny overseas. So under what circumstances would the activities of U.S. citizens be revealed to foreigners; the British, for instance, had justified complaints that U.S. citizens were aiding Irish Republican Army (IRA) guerillas with donations. Can the United States trust that other nations would adhere to similar standards of data confidentiality and appropriateness? Perhaps with like-minded countries, this is a minor issue. Indeed, Europeans complain that too much data are collected in the United States. Conversely, certain speech acts are crimes in France and Germany but are considered protected speech here, and thus the United States would be disinclined to collect or share files on what people said.

Whose mores apply in cyberspace may be strongly influenced by who gets a functioning system up first. Those who do will have economics on their side; if, for instance, system A and system B are equally costly when a country is starting afresh, but one country builds system A first, the second country will usually find that system A is easier to build than system B. In addition, both countries will have a leg up in having their protocols, procedures, and data mining algorithms accepted globally. They will also have a starter set of initial identifications and correlated events that can be used for feeding, calibrating, and testing everyone else's system.

Whichever biometrics are selected to ascertain identity – seemingly a mere technical matter – can have broad ramifications. As noted, if the purpose of an international identity system is to determine who is trying to assume two identities, it is important that everyone use the same biometrics. If fingerprints are used, people can be vetted for entry or treatment by their history of crime. If faces are used, surveillance data are more likely to be appended to checkpoint events. If DNA is used, it is easier to build and interpret familial networks. If irises are used, it is more likely that none of these ancillary features will be exploited.

The choice as to what to collect and how to analyze it (for example, to determine how suspicious certain individuals are) may also influence others to follow suit. If the United States collects information on certain events, other countries that might want to collect information on other events may find that, when finished, they have only a partial database;

had they chosen the same events as the United States did, they would enjoy a more complete database. The more that analytic tools are useful, globally consistent (that is, behavior that is noteworthy in one country is noteworthy in another), and require collecting information on certain checkpoint events and not others, the greater the incentive of other countries to adopt them. The more they embed certain assumptions about such events (such as which events trigger what responses and correspondingly develop similar checkpoints), the greater the influence of their adoption.

Even something as basic as a watch list has such ramifications: On what criteria is it based? Can such criteria be shared? Are the risk criteria used in building such a list consistent? A watch list that errs on the side of suspicion will be larger than one that errs on the side of proof.

As with the NATO case of the last chapter, the ability to take a lead in cyberspace in this day of expensive systems integration creates the likely possibility that other countries will add their monitors and clients to the dominant system rather than build their own. Whether they decided to do this will, admittedly, depend on other factors, such as whether the system is a sufficiently good fit to how other countries do their business and, in practice, how much business local firms get from the deal.

7.3 Compare, Contrast, and Conclude

Giving something away without asking for something specific back is not always the obvious course of wisdom in business or public policy. But, since the cost of duplicating information is low, the willingness to give others access to information and information services can be easily used as a quid to bargain for someone else's quo as a way to foster interdependence or even asymmetric dependence, and to influence how sectors and activities evolve.

Both geospatial data and identification data arise in large part from surveillance systems. Both are largely motivated by security. Both, therefore, are the province of organizations that like to keep secrets. The latter, of course, means that the manipulation of cyberspace along the lines argued in this chapter is neither natural nor therefore inevitable.

The more interesting differences between geospatial and identification domains lie in the importance of interpretation in wresting value from data. Geospatial data are all about interpretation for natural (for example, what is a river, a stream, a lake) and especially manmade features (what is a campus, for example). Biometric data are straightforward: these two people either are or are not identical. Yet, when identification data and checkpoint data are married, the latter are subject to considerable interpretation. Interpretation requires semantics; semantics, in turn, requires categorization, and categorization influences thought.

8

Retail Conquest in Cyberspace

Espionage and propaganda are two pillars of traditional information operations. Espionage ferrets out small amounts of valuable data (for example, to determine whether a particular country has a nuclear weapons program). Propaganda broadcasts messages to alter another nation's perceptions and hence decisions. Both operate at the national – that is, wholesale – level, even if the minds they would steal from or change are those of discrete individuals.

Exploiting cyberspace enables information operations at the retail level. Unique information can be gathered person by person from a large number of individuals, and messages can be tailored to every listener. Such conquest involves a mix of friendly and hostile approaches. The friendly part is inducing people to reveal information or get them to listen to your story. The hostile part is stealing information from others or using such information to coerce others.

This chapter discusses conquest at the retail level. Following some scaffolding, it examines how the necessary information is acquired and then used. The section on information acquisition is divided into surveillance in real space, surveillance in cyberspace, the transfer of information from the local to the global level, and privacy issues. The section on information exploitation discusses how such data can be amalgamated and exploited, but ultimately what such data are worth.

8.1 Information Trunks and Leaves

Although the sizeable U.S. and Soviet intelligence systems built during the Cold War collected reams of data, they (or at least ours) did so to inform ourselves about a few key facts: what were the other side's intentions, what weapons did they possess, what were their force deployment policies, and who their spies were. Such data could be expressed in a few bytes even if it took a huge dredging operation to identify the nuggets with sufficient confidence.

Likewise for the information we ladled on them. Broadcast information (whether via Radio Free Europe or via public speeches) can convey only so many bits of data. Rarely did such information differ in content, translation nuances aside, from one subset of the population to the other. In both cases, the message was composed of small amounts of important information.

When every bit matters, each is handled with care. Everything is checked and rechecked. Data are stored in many different places and constantly compared with associated data; for example, does the premier's meeting agenda comport with his travel schedule? Many people keep such important data topmost in their minds. If our knowledge of their secrets is itself secret, then few people get to know, their propensity to leak is carefully assessed, and the handling of information is subject to arcane rules. Some people spend their lives looking for evidence that such information may have already leaked. Similarly, if there is only one message to get across, great thought will go into how it is packaged and sold.

The information revolution – that which supposedly justifies the newfound importance of information warfare – has largely been felt at the low end, though.[1] As noted in Chapter 5, the enormous growth taking place is of low-grade data: more of it is being collected, and, thanks to networking, more of what is collected locally can be accessed globally. After all, high-value information has been worth amassing, manipulating, and transmitting even in an era when the cost of a bit was not as trivial as it

---

[1] An analogy may be made to the impact of technology on the supply of copper and gold. Very few high-grade deposits of such metals are being discovered these days. However, the industry's ability to make money on low-grade deposits keeps getting better.

is today. Low-value information was ignored if it was too expensive to analyze or use. As costs declined, successively lower-yield information crossed the threshold to where it was worth collecting, transmitting, and storing.

Admittedly, some very important pieces of information (such as WalMart's annual revenues) are composed of, say, billions of individual transactions. Sometimes, important and unimportant data are simply two different classes of information. It is vital that the U.S. Navy know the location of each of its hundreds of ships. It is also important that it know the location of each of its several hundred thousand sailors, but each sailor's location is individually less important. The latter is collected and processed in a different manner.

There is a similar, albeit less direct, relationship between the small and the large in persuasion. These days, one must influence the mass of individuals to be sure of influencing states in the long term.[2] Politically, the mass of individuals matter in democratic states. Economically, they do in states with a sufficiently broad distribution of discretionary income. Militarily, good armies rely on the willing and enthusiastic exertion of force by everyone in arms; rote obedience is incompatible with the tools of modern warfare. And even undemocratic states with little discretionary income have to be sensitive to the street. With terrorism, otherwise irrelevant and hitherto anonymous people can, through their choice of occupation, exercise disproportionate influence on a state's behavior and well-being.

Psychological operations, in other words, will have scope to evolve from a wholesale to a retail trade.

8.2 Where Does Cheap Information Come From?

It is easy to show how the information revolution has magnified the quantities of low-end information in circulation.

---

[2] The information technologies of the last quarter-century have generally tended to magnify the power of individuals relative to that of states. For an early argument as to why this is the case, see, for instance, Ithiel de Sola Pool, *Technologies of Freedom*, Cambridge (Belknap Press), 1983. As explained in Chapter 6, the proliferation of cell phones and the Internet have only furthered this tendency.

Because most purchases, for instance, are made via bank cards, they are all recorded somewhere. In the last ten years, thanks to the potential power of customer relationship management (CRM) software, most large food and drug chains have issued affinity cards. Issuers of these cards, or any business-to-consumer E-tailer, not only know how much you spent when, but on what; this permits them to offer targeted "deals" on particular items. The amount of information required to fill out government forms only goes up – especially in schools, where no activity takes place without being preceded by a blizzard of permission slips. Medical release forms are no exception either. There is no shortage of transactions waiting to be amassed, amalgamated, and analyzed.[3]

Surveillance devices are generating new forms of data.[4] Webcams (cameras whose video goes simultaneously to the Web) are already popular devices in such places as congested thoroughfares or the panda cage at the National Zoo in Washington, D.C. The heightened sensitivity to terrorism after the World Trade Center attack has also resulted in increases in surveillance. By that point, the UK already had two million video cameras[5]; they provided copious shots of the July 7, 2005, suicide bombers after the event. The software to recognize faces and thus identities from such sightings will permit an increasing translation of data such as facial images into information, such as whether a particular person was at a specific location at a precise time. A greater propensity to ask people to sign (manually and, increasingly, electronically) into hitherto less-sensitive spaces is likely to augment a large source of attendance data. Automated toll-collection devices, which are associated with most well-traveled toll roads in this country, can track the passage of vehicles – and thus their probable drivers. Many manual toll collection points (such as airport

---

[3]   See, for instance, Michael Froomkin, "The Death of Privacy" *Stanford Law Review*, 52, May 2000, p. 1461–1543; Daniel Solove, "Digital Dossiers and the Dissipation of Fourth Amendment Privacy," *Southern California Law Review*, 75, 2000, pp. 1083–1167.

[4]   See, for instance, Robert O'Harrow, *No Place to Hide: Behind the Scenes of Our Emerging Surveillance Society*, New York (Free Press), 2005; Simson Garfinkel, *Database Nation: The Death of Privacy in the 21st Century*, Sebastopol (O'Reilly), 2000; Jean Kumagai and Steven Cherry, "Sensors and Sensibility" and subsequent articles, *IEEE Spectrum*, July 2004, pp. 22–48.

[5]   See Jeffrey Rosen, "A Watchful State," *New York Times Magazine*, October 7, 2001, p. 38.

parking lots) take snapshots of license plates. With software of no great sophistication, they can be read, correlated, and stored.

The global population of sensors is growing and becoming networked. They range from everyday devices as cameras, microphones, and motion detectors, to electronic RFID readers. Outdoor versions can monitor the environment and detect biological or chemical agents. Internal sensors can continuously read health indicators for at-risk patients. Networking reduces the costs of monitoring and harvesting information from devices, especially those far away or embedded in something else. It permits data to be analyzed as they are collected. For this one must thank the declining cost of the devices' economics, increases in the sensitivity of engineered materials, and the coming of microelectromechanical (MEMs) devices (such as those used in air bags).

Coupling sensors with large databases may allow authorities to recognize people they have never met when they are encountered in public places. Football fans with tickets to the 2001 Super Bowl in Tampa had their faces scanned by systems that tried (and mostly failed) to match their facial attributes to a database of known criminals and troublemakers.[6]

Added to that are several emerging sources of location data for people or at least their devices. At least one car rental company has put transponders on its rental cars to determine which were being driven in excess of 75 miles an hour.[7] Cell phones periodically chirp to tell cells where they are. When cells shrink (for example, by going to the smaller Sprint PCS cells), locations determined by triangulating from signal strength at each cell have that much less error. U.S. cell phones are further evolving into GPS devices so that the source of enhanced 911 (E911) calls can be determined to within 20 meters. The RFID-based proximity card is rapidly becoming

---

[6] The efficacy of the facial recognition system was admittedly low (see John D. Woodward, *Super Bowl Surveillance: Facing Up to Biometrics*, Santa Monica [RAND], 2001). Face recognition works best among small groups or when there is some a priori reason to believe that someone specific may be in front of the camera. There may be limits to how far technology can improve; the human face, as measured in two dimensions, may be insufficiently stable and unique (in other words, two faces at certain times can look too similar to be distinguished).

[7] See Robert Lemos, "Rental-Car Firm Exceeding the Privacy Limit?" http://news.com.com/2100-1040-268747.html, June 20, 2001. Incidentally, Acme Rent-a-Car lost a court case where the legality of the practice was challenged.

standard in any institutional environment where physical security is a concern.[8] Although such cards are generally used to determine whether someone can enter a facility, such data can and often are centrally collected.[9] RFID-based electronic bar-coding[10] appended to merchandise will let it reply, transponderlike, to broadcast queries, thereby revealing its location.[11] Initially only expensive goods and packages would receive such treatment; as costs drop, the bar codes can be affixed to razor blades and such.[12]

Where boxes lead, humans may be dragged into following. People in institutional settings such as hospitals are still poorly tracked, particularly when transferred from one institution to another. Such people may one day be effectively bar-coded for electronic tracking. Indeed it is becoming far less expensive to monitor the vital signs of patients, keep them at home, and respond to emergencies than it is to hospitalize them[13] – and electronics is getting cheaper and less power-hungry while networks, especially RF networks, are becoming more ubiquitous.

8.3 Surveillance in Cyberspace

Personal data are also being collected directly through cyberspace. The rise of Web surfing has meant an enormous increase in information on what people look at, when they look at it, and how long they spend

---

[8] As Exxon's Fastpass suggests, proximity devices can be used in lieu of credit cards to record purchases.

[9] See Edward Balkovich, Tora K. Bikson, and Gordon Bitko, *9 to 5: Do You Know If Your Boss Knows Where You Are?: Case Studies of Radio Frequency Identification Usage in the Workplace*, TR-197, Santa Monica (RAND), 2005.

[10] For a fuller treatment of the issue, see Simon Garfinkel and Beth Rosenberg, ed., *RFI: Applications, Security, and Privacy*, Upper Saddle River (Addison-Wesley), 2005.

[11] Winston Chai, "Radio ID Chips May Track Banknotes," http://news.com.com/2100-1017-1009155.html, May 22, 2003, referring to a reported deal between the European Central Bank and Hitachi.

[12] The ability to exchange data with satellites requires fancier electronics and more battery power than the ability to do so within, say, a warehouse. The use of satellites may thus be merited only for vehicles, whereas the lesser technology may be applied to boxes in a warehouse.

[13] One hospital chain located in Virginia's Tidewater, Sentara Healthcare, already uses electronic monitoring to keep a constant track of intensive care unit (ICU) patients; see David Brown, "Intensive Care, from a Distance," *Washington Post*, 125, 179, June 2, 2002, p. A1.

doing so. Many users routinely trade personal information for Web-based services. Some use calendaring functions that certain Web sites supply for free. TiVo,[14] a company that sells devices to time-delay television shows, has a premium subscription service that would have users reveal their television viewing habits so that their shows can be time-sequenced more to their liking.[15]

With its now-scaled-back Passport[16] and scuttled Hailstorm initiatives, Microsoft would have gone a step further and used the Internet to acquire personal information about the service's users. Passport is essentially a one-stop shop for password access and E-commerce information (including credit cards, shipping addresses, and ancillary preferences). Rather than remembering a blizzard of passwords, the theory goes, a user would simply log into Passport, which would, in turn, pass such information on to each of several Web sites that users may want to visit, for a small percentage take of any subsequent transactions that result. Early reports of Passport have noted its potential to lead to privacy violations.[17] Others see security vulnerabilities multiplied by putting all one's password eggs in the same basket.[18] Hailstorm, for its part, was to be something like

---

[14] See "Nielsen Begins Monitoring TiVo Usage," http://news.com.com/2100-1040-948580.html, August 6. 2002.

[15] In late April 2002, a coalition of media companies persuaded a judge to order SonicBlue to turn over records detailing which shows users watch, when, and whether they skipped commercials while watching them. Christopher Stern, "Privacy Fight Centers on Ad-Zapper," *Washington Post*, May 4, 2002, p. A1, A12.

[16] In 2004, Microsoft recast its .Net Passport identification system, limiting the service to its own online offerings and those of close partners. Microsoft no longer saw Passport as a single sign-on system for the Web at large, a spokesperson said. See Joris Evers, "Microsoft Scales Back Passport Ambitions," IDG News Service, www.infoworld.com/article/04/10/20/HNmsppscaleback_1.html, October 20, 2004.

[17] Brian Livingston reports on a finding by Bugtoaster.com that Windows 9x computers infected by Trojan horses can easily steal Passport's passwords and thus credit card information. Brian Livingston, *Information World*, October 1, 2001, p. 46. In early August 2002, Microsoft and the Federal Trade Commission (FTC) reached a settlement on the privacy issue following a summer 2001 complaint filed by thirteen consumer groups. They alleged that Microsoft made it "difficult if not practicable" for consumers to exercise control over their personal information. See also Wayne Rash, "Your Stolen Passport," techupdate.zdnet.com/techupdate/stories/main/0,14179,2814881,00.html, September 26, 2001.

[18] Microsoft was forced to suspend an important financial service of Passport when a programmer showed that he could obtain such credit card numbers merely by sending victims a single e-mail. Brian Livingston, *Information World*, November 19, 2001, p. 46.

a life-minder.[19] Users would give Microsoft their calendars, schedules, various contact numbers, contact lists, and so on. So, for instance, if your flight were scheduled to depart late, Microsoft would alert you by phone, beeper, or e-mail. If your flight arrived late, Microsoft would calculate which meetings you would likely to be late for or miss and who among the contact lists needs to know – and it will inform them. One early service, rolled out in 2001, used the company's instant messaging service to send bulletins on stock updates, auction bidding, and so on.[20] Whether users will feel comfortable telling Microsoft what they otherwise would have entrusted to their secretaries is unclear. But, if it becomes popular, someone in cyberspace will know a great deal about users. Although neither Passport nor Hailstorm is showing much life, the impetus that led to their introduction still exists. Google, for instance, offers its G-mail account in the hopes that users will let Google comb through their e-mail as a way of letting Google target advertisements to them.

Mere appearance on the Web, however, does not necessarily identify one because cross-walking identities from cyberspace to real space is not trivial (even if 90 percent of all IP addresses can be linked to an

---

[19] Reports Steve Gilmor:

> Hailstorm was [sent back] to the drawing board in 2002 [because] Hailstorm was started before XML query and other important XML Web services standards would be finished. So Mark Lucovky and his team had to "roll their own" XML technologies which now need to be replaced.... [Said Bill] Gates, "our strategy that everything we do should be common between how it works on a rich client that works offline; how it works on a server where a corporation can come in and license the software and run that themselves – and how it works on a service where somebody who doesn't want to run that server can connect to Microsoft and one of our partners and under some sort of economic structure, most likely a subscription like structure, have access to the capabilities of that service running out in the Internet" (Steve Gilmor, *Infoworld*, August 5, 2002, p. 50).

> Or as Tom Yager has speculated: "Companies that want to host Hailstorm internally can buy the server instead of struggling with issues of privacy, control, security, and billing accuracy raised by connecting to a Microsoft-hosted service." Tom Yager, "Microsoft Rethink," http://www.infoworld.com/article/02/04/26/020429opestrat_1.html, April 9, 2002. A third problem is that Hailstorm threatened to place Microsoft in a position to compete with its customers. Brian Fonseca and Ed Scannel, "Microsoft Plans XML Politics," *Infoworld*, April 15, 2002, 4.15.02.1.

[20] Mike Ricciuti and Robert Lemos, "Network: Reinventing the Wheel in Real Time," http://news.com.com/2009-1001-274440.html, October 22, 2001.

approximate physical location).[21] Multiple family members may still be using the same mailbox.[22] One individual could use multiple aliases – a standard feature on Windows XP. Packets that flow to Web hosts normally indicate only the IP-address of the viewer. These days, with the Internet's free address space growing scarce, most users have dynamically allocated addresses. It is next to impossible to determine, absent ISP help, anything meaningful from such addresses. However, identifying information is often found in a user's files, most notoriously the cookie.txt files invented by Netscape in 1995.[23] If and when IPv6 (Internet Protocol version 6), with its enormous name space, is implemented, everyone can go back to having a unique IP address, simplifying the correlation.[24]

---

[21] But this is a problem companies have been, controversially, working to solve. In mid-2000, Cogit, a Web company, began matching user visits to Web sites against Polk's database of 110 million names, addresses, and "lifestyle characteristics." Brian Livingston, "New Web Tracking Service Raises Old Privacy Concerns," http://news.com.com/2010-1071-281326.html, June 16, 2000. This matching begins whenever a person at a participating site reveals a name, address, or other identifying information; a match creates a profile of the person's demographic information (income, net worth, bankruptcies, home value, cars owned, religion, ethnic group, investments, and so on). A unique serial number for the profile is created and then stored as a cookie on the consumer's browser, the policy says. This followed DoubleClick's acquisition, for similar purposes, of Abacus Direct, an offline marketing company.

[22] This seems likely to change, however. AOL's e-mail policies, "free" e-mail Web services, and the Windows XP operating system suggest that everyone will one day get their own.

[23] A cookie.txt file is used to hold information generated by the Web site itself. It was invented to identify a specific user from one part of a transaction (such as finding a book) to the next (putting in a "shopping cart"). This file enables a vendor, such as Amazon, to display a user's "shopping cart" in response to an HyperText Transfer Protocol (HTTP) request that would otherwise only display a meaningless IP address. See, for instance, John Schwartz, "Giving Web a Memory Cost Its Users Privacy," *New York Times*, September 4, 2001, p. A1.

[24] The first live implementations were reported in early 2002. Stephen Lawson, "IPv6 Enters the Real World," *Infoworld*, February 2, 2002, p. 35. Much of the impetus for IPv6 comes from countries that are now large Internet users but accounted for a much smaller share of the Internet at the time that address blocks were first being allocated. The federal government is hoping to shift all of its computers to run on the next-generation Internet by 2008 and has developed contracting vehicles and funding streams accordingly. Dawn Kopecki, "Where Opportunity Is Knocking for Small Business," *Business Week*, December 4, 2006, p. 81. As of 2005, however, only DoD seemed to be ready for the switch. See Anne Brache, "Feds Slacking in Shift to Next Generation" http://news.com.com/ /2100-1028_3-5768937.html, June 29, 2005.

There are other ways available today to link virtual and real identities. Some people routinely reveal their real identities in their e-mail via signature blocks. Information that relates e-mail addresses to real people may be purloined from sloppy ISPs. Many merchants couple real-space information (such as names on credit cards and shipping addresses) and e-mail information (from correspondence, for example). Data consolidators such as Acixom have many tricks for consolidating identities by examining patterns of real-space transactions; such skills may be applied to resolve identity questions in cyberspace. People, especially bloggers, who have written and made public enough material under their own name will have established grammatical and word choice patterns that may be discerned in their e-mail or instant messaging, allowing real identities to be inferred from the latter. Nondestructive worms or viruses loaded into a user's machine may reveal a correspondence between information in files and the user's surfing habits (as such, these worms and viruses serve as a kind of super-cookie).[25]

One need not be a vendor to gather personal information. One can establish a conversation (through IM, for example) with people in cyberspace and, from there, cajole or trick them into revealing such information. A small but sophisticated coterie of friends may be able to work chat rooms, Usenet sites, or instant messaging cells that may be a vehicle for eliciting opinions on a broad variety of topics much as focus groups do, providing raw material from which to build advertising and propaganda campaigns. Greater harvests may be achieved by developing a silicon conversationalist that can pass a Turing test to the point where users have little clue that they are not talking to a human.[26] Those who can do this well (and can link virtual personae and real persons) might be able to gather profiles on millions of people with no more energy than labor-intensive methods could previously have achieved for thousands. The tens of millions of bloggers around the world working the boards today may not even have to be worked hard; they reveal themselves for free.

---

[25] For a discussion of what some of these viruses are and what they do, see Chris Taylor, "What Spies Beneath," *Time*, October 7, 2002, p. 106. See also John Borland, "Spike in 'Spyware' Accelerates Arms Race," http://news.com.com/2009-1023-985524.html, February 24, 2003.

[26] Sherry Turkle, *Life on the Screen*, New York (Simon and Schuster, 1995), pp. 25, 88–96.

8.4 Making Information Global

The transition from local to global information is one between some-one knowing some information and everyone potentially knowing all information.

Several factors, notably digitization and networking, are at work to promote the globalization of information.

Digitization makes storage of communications the norm and erasure the exception. Analog phone lines, for instance, have to be specifically instrumented if the signals they transmit are to be captured; by con-trast, it is in the nature of packet switching that bits are constantly being received, copied, and then retransmitted.[27] Copying them one more time for permanent storage is no big deal. In the past, multiparty conversations often had to take place in person because conference-calling remains a difficult-to-work feature that is rarely used by home callers; today such a conversation might be mediated through chat rooms or, in years to come, via PC-based videoteleconferencing. Keystroke monitors on computers can tell whether employees are attentive to their work – but they can also serve as a more permanent surveillance mechanism.

In other cases, however, one must explicitly decide to store information even if storage is easy to do. A proximity card informs a local decision – whether or not to open a door – and usually within a second; thereafter, the information on which the decision is made loses its *primary* value. But has it lost its *secondary* value? Does someone want a permanent record of everyone's comings and goings?[28] Well, they might want evidence that employees are engaging in suspicious activity, or maybe just goofing off. They may want to know where employees are in case of trouble. And, they may have benign reasons, such as figuring out how to lay out a building to minimize unnecessary foot travel. Conversely, there

---

[27] For instance, in analog systems, messages left on message machines are usually erased when heard and usually inaccessible to those without physical access to the devices themselves (unless they can guess the two-digit security code). Systems that integrate voice mail and e-mail give voice messages the lifetimes of e-mail messages; ensuring that all copies are erased can be very difficult.

[28] Proximity cards are frequently encoded with information that identifies permanent holders such as employees. Some institutions) also use them for visitors, but not in ways that allow *specific* visitors to be tracked.

are sound reasons *not* to collect such information. It signals distrust of employees. Unscrupulous monitors can abuse such information. And, once recorded, such data can be subpoenaed. But the latter are secondary considerations.

Costs, while declining, are not necessarily trivial. To reengineer a local device to feed a global database requires connectivity – and thus either wiring, some wireless connection, or at least onboard storage for later pickup. Wiring means tearing up or at least affixing something to walls. Wireless connections require bandwidth and security unless someone's eavesdropping or hijacking such connections is deemed unlikely or unimportant. Building and managing network connections as well as setting aside storage and processing capacity are also necessary. Collecting records from unconnected devices entails traveling from one to another periodically. Someday, every sensor may carry within it low-power wireless connectivity to feed networks to which they report – but not yet.

Signals collection, nevertheless, is burgeoning.

8.5 Privacy

Is Scott McNealy correct when he argues that "you have no privacy, get over it"? Are there no political roadblocks to the naked future?[29] Polls consistently show that people feel they have too little privacy on the Internet, but whether they do something about it is another matter. Paul Saffo has a point when he argues, "Americans talk about privacy, but they'll spill their guts for a package of trinkets."[30] Few folks actively manage their cookie.txt file to hide their virtual travels.

Not that managing privacy has been made easy. Many Web sites have posted privacy policies on their site – but often off on its own page, not necessarily worded for clarity, and thus rarely consulted by Web surfers.[31]

---

[29] Steven Hetcher, "Changing the Social Meaning of Privacy in Cyberspace," 15 *Harvard Journal of Law and Technology*, 2001, pp. 149.

[30] Quoted in Bob Tedeschi, "Privacy vs. Profits," http://www.findarticles.com/p/articles/mi_zdzsb/is_200110/ai_ziff14753, October 2001.

[31] This is also true in the real world. See Mike France, "Why Privacy Notices Are a Sham," *Business Week*, June 18, 2001, pp. 82–3. Circa 1999, the *Georgetown Internet Privacy Policy Survey: Report to the FTC*, June 1999, reported that among sites that

One standard, the Worldwide Web Consortium's (W3C) Platform for Privacy Preferences (P3P),[32] would let users declare what information they will and will not permit to be transferred to Web sites and under what circumstances – without their having to read each site's policy language. Negotiations between each user's policies and Web site policies would, in theory, permit only certain information to be passed. Microsoft's Internet Explorer version 6 supports P3P. Yet, users have to know how to use the feature; if not set, it blocks nothing. Sites have to know how to use the feature as well. They also have to be trusted to adhere to their policies. Interest in P3P among Web site owners seemed to have died down circa 2003.

Third-party institutions such as TRUSTe exist to certify the privacy policies stated on Web sites; many sites carry TRUSTe's approval seal.[33] Nevertheless, site owners could be devious and clever about fooling authenticators or willing to collect information now and violate its self-declared disclosure strictures later, especially if they are sold to another firm with no such qualms. During the dot-com bubble, the market valuation of many firms (such as Yahoo's one-time $150 billion market capitalization) was said to be based on the financial advantage of possessing a great deal of customer information. Complete customer profiles were valued as high as $1,000 each; one plucky individual even offered to sell

---

collected personal information, two-thirds had a privacy policy in place. According to J. Adkinson, F. William, J. A. Eisenach, et al., *Privacy Online: A Report on Information Practices and Policies of Commercial Websites*, Washington (Progress and Freedom Foundation), 2002:

- As of late 2001, Web sites were collecting noticeably less information than two years earlier.
- Fewer Web sites use third-party cookies.
- Privacy notices are more prevalent and prominent.
- Opt-in policies are gaining in popularity; opt-out policies declining.

[32] Interest in P3P started declining in 2002, when major sites proved reluctant to implement it. Paul Festa, "Promise of P3P Stalls as Backers Regroup," http://news.com.com/2100-1023-963632.html, October 29, 2002.

[33] TRUSTe's credibility was called into question when it refused to revoke its certification of Real Networks, whose RealJukebox application "had been distributing software that surreptitiously gathered personal data from users' hard disks. TRUSTe responded by claiming such actions were beyond the scope of the TRUSTe audit because the RealJukebox software worked only indirectly through a Web site visit." Batya Friedman, Peter Kahn, and Daniel Howe, "Trust Online," *CACM*, December 2000, pp. 34–40.

his dossier for that much, albeit unsuccessfully. Today's valuations appear more realistic, that is, lower. Still, a good customer database has value.

Unfortunately for the clarity of discussion, the notion of privacy subsumes two similar but separate desires. One is to keep hidden that which is of a private nature (such as medical records and religious confessions). The other is to prevent information resulting from a disaggregated local transaction from being amalgamated and made global – and thus accessible to unknown others. In the United States at least, the first is far better protected under law and the Constitution. Europe, by contrast, has asserted a broad privacy right by enacting a comprehensive directive that forbids information collected for one purpose to be converted into another. Such strictures have had some effect on the activities of multinationals with reputations to consider and legal staffs to keep happy. But in cyberspace, a type of Gresham's law rules. The badly behaved can sit wherever laws are lamest.[34] If working beyond the law's reach provides a competitive advantage, then the lax will outpoint the law.

What is it about the globalization of personal information, however, that does make it a privacy problem? Perhaps what can be revealed to a stranger (such as a drugstore clerk) cannot, without consequences, be revealed to someone with power (such as one's boss). Perhaps what is benign when separate (for example, being seen purchasing a prophylactic and knowledge that one's wife is past menopause) becomes damning when combined. But maybe something else is at issue.

8.6 Amalgamating Private Information

The existence of personal information everywhere does not mean the existence of personal information in any one repository – a prerequisite for understanding one individual well enough to determine to which appeals he or she might respond. So, (1) how can private information can be amalgamated, and (2) how could information warriors exploit it?

---

[34] True, the European Union (EU) prevents those collecting information within member countries from transferring them to countries with less restrictive privacy regimes (unless, as with many U.S. multinationals, they have made specific "safe harbor" arrangements to protect the data). Yet, an EU citizen going to a Web site hosted outside the EU (something very difficult to determine by inspection) has no such protection.

Amalgamators could:[35]

- Collect the information themselves through physical or cyber surveillance
- Persuade people to yield such information voluntarily
- Purchase it – licitly or illicitly – from those who have collected it
- Steal such information from one of the following sources:
  - Those who have collected it
  - Users
  - Networks (through taps, for example)

Physical surveillance has the advantage of not needing anyone's permission or, quite often, knowledge. But surveillance of a population large enough to interest information warriors is likely to be not only complex but expensive as well. The United States is a big country.

Persuasion, especially through the use of smart software capable of carrying an interesting conversation, requires less equipment, but such software has yet to be written. Such a stratagem would unearth the secrets of those who would reveal themselves to complete strangers in cyberspace – and how many people would do *that* is unclear.

Purchasing such information has its attractions. Many consolidators already collect personal information, although most of what they purchase is already in the public domain (such as real property records, court judgments, and arrest records). Although they profess that they have no interest in selling personal information to criminals,[36] it is hard to believe that they would not sell to those interested in propaganda more than they are in merchandising. A company that promises to be a privacy hawk may also be bought by a privacy snake with no intention of honoring such promises. State governments have, controversially, sold drivers' license lists. Nevertheless, many of the larger collectors of information over the Internet (including Microsoft) have broader public trusts to maintain and are thus less likely to sell personal profiles – nor is bankruptcy all that likely for them. The next best option there is to find employees willing

---

[35] If the amalgamator is a government, it has a further choice of asking for the information under color of law (for example, through a subpoena) – but that introduces a completely different set of issues.

[36] But they do anyway, or at least did. In early 2005, ChoicePoint mistakenly sold personal credit reports for about 145,000 Americans to criminals.

to leak – perhaps one in a hundred can be bought. Foreign governments that protect information on their own citizens may have fewer qualms about turning over information on others.

How likely is it that hackers can come up with such information?[37] As noted, otherwise benign viruses or worms may be able to steal personal information from home computers and quickly disappear, but they would need to be intelligent to avoid having to export every document they see – in which case the unusual volume of outgoing bytes may arouse suspicion.[38] Putting sniffers on enough Internet gateways requires subverting network operators, a breed more security-aware than most mere users are. Hacking into databases directly may work spectacularly against some of them, and the information so purloined can fuel identity theft. Such hacking, though, may not necessarily reveal enough information to build up a sufficiently indicative profile of any given individual.

8.7 Using the Information

Assume success, despite obstacles. Tomorrow's warriors in cyberspace *do* manage to collect all this information. What could they do with it?

8.7.1 General Coercion

Start with the paranoid fantasies. An information warrior learns details about you – maybe financial and medical secrets. Perhaps this someone does not know the darkest details, but reveals enough of what you thought safely hidden to suggest that a great deal more is known. As the blackmail victim, you are simultaneously threatened by exposure and unable to turn to others. The price of your safety is not great: a task here, a word

---

[37] The year 2005 saw a rash of data-theft discoveries. On August 12, 2005, for instance, a Florida man who ran a bulk e-mail company was convicted of stealing more than 1.5 billion data files from Acxiom. See http://www.consumeraffairs.com/news04/2005/acxiom_levine.html.

[38] The FBI's Magic Lantern project reportedly lets it plant a Trojan horse keystroke logger on a target's PC by sending him or her a virus over the Internet rather than requiring physical access to the computer as is now the case. See "FBI Confirms 'Magic Lantern' Project," www.cnn.com/2001/TECH/internet/12/13/magic.lantern.reut/index. Firewalls that detect unusual outgoing traffic may pick up exploitation earlier than otherwise.

there, a little more information collected on your friends. The further the dealings with this mysterious someone go, the more you have to be embarrassed about; information you reveal about others, in turn, makes them more vulnerable and makes you feel more guilty. Before long, you are willing to do anything to avoid revelation, making you a pawn in their game. Voilà: third-rate screenwriting.

True, there have to be some people whose background leaves them vulnerable to such blackmail, but whether they can be identified from physical and cybersurveillance is something else entirely. More commonly one is working with the accretion of the everyday. People may not want it known that they have, say, a fondness for tuna fish as unearthed by the records of their affinity card usage. Yet, diffidence is not an adequate basis for blackmail. The Victorian novels that turned on efforts to keep or expose deep dark secrets presupposed a society in which shameful secrets actually existed. Perhaps today's Americans yield their privacy so readily because, in today's "whatever" culture, stigma has lost its sting. The chill that most modern folks experience from being naked may well be more physiological than psychological.

Statistically speaking, unless the few who are vulnerable to blackmail based on easily collected information have links to one another (rather being than spread randomly throughout the population), the chances that subverting one can lead to subverting many have to be low. There may simply be no critical mass. If one cannot identify such people beforehand, blackmailers run the risk of targeting people who (1) do not care and (2) have no problems alerting authorities.

This lurid world of suspicion and betrayal is redolent of 1984's "I sold you and you sold me." Yet, it had a real-life counterpart in East Germany's Stasi, which worked hard to make mutual betrayal under pressure a sad fact of life. Stasi, though, enjoyed the machinery of economic and physical repression as well as embarrassment. When the tide turned against the East German regime in 1989, it collapsed as fast as did sister Warsaw Bloc countries with less efficient secret services.

8.7.2 Specific Coercion

If used against more influential targets, could a coercion campaign work better and cost less to collect information for?

Politicians, for instance, are clearly more influential than common folk and should be good targets. Certainly the Hoover-era FBI held volumes of dirt on many of them for *some* purpose. But the world has changed since then. Those who love their privacy have absolutely no place on a campaign trail. Conversely, the voters' tolerance for bad personal behavior is considerably broader than it used to be (whereas the tolerance of politically unpopular wisecracks is not what it was). If it has yet to descend to the level perceived by former Louisiana Governor Edwin Edwards[39] – it is getting closer.

Senior civil servants, at least the majority who have never run for election, may have received less public scrutiny. Having more privacy to preserve, they may make better targets. But those who fill national security positions are subject to increasingly detailed surveys of their personal lives for security clearances. The tribulations of Zoe Baird, who lost her bid to become Clinton's attorney general when controversy erupted about her hiring of an illegal alien, only accelerated the intrusiveness of the preclearance process for high office. Senior military officials face similar security screening even if public scrutiny is generally less intrusive (the treatment of General Joseph Ralston during his consideration for the chair of the Joint Chiefs of Staff came close, though). Military service affords less privacy in their lives, and their peers, before whom they stand, adhere to fairly conservative behavioral norms. All this limits the potential for coercion, even if military personnel are, ultimately, human. Yet, dedication to the public cause and the pride they would lose should their actions be seen as influenced by coercion nullify many benefits of collecting information on them.

Threats against fielded forces may have to be taken more seriously.[40] World War II propaganda tried to play on the anxiety warfighters normally feel over their vulnerable wives and children back home. The

---

[39] As per his quote, "I'll keep getting elected as long as they don't catch me in bed with a live boy or a dead girl." *Time*, March 11, 1985, available online at http://www.bartleby.com/63/35/635.html.

[40] Incredibly, the DoD considers the unit assignments of all former servicepeople to be public information. Such data were released in toto to the Web site Military.com. The potential for intelligence collection and concern inherent in such a policy can only be imagined.

ability to collect information on individuals, however, could make such threats more specific. Indeed, there is indication that Serbians knew and exploited personal information about specific airmen during the 1999 Kosovo campaign; earlier threats were reported in Desert Storm. Such efforts availed naught, in part because Serbia lacked a terrorist cohort in the United States, but the same cannot be said about current foes. That noted, one hardly needs a great deal of personal information on families back home to worry warfighters; a well-publicized crime against a random victim will raise everyone's anxiety levels.[41] Specific information is only needed to influence specific warfighters and only if their acts can be credibly linked to any harm that befalls their families. But that requires knowing, in detail, who has done what on the battlefield – which normally requires far more situational awareness than U.S. foes possess.

8.7.3 Persuasion

Imagine seeing files on you held by a foreign country. These files appear at first glance to be accurate. Your name, address, birth date, occupation, and when you entered that country are correct. More data are presented. The total purports to be a complete profile of you. You grow upset, and increasingly worried; you feel vulnerable, almost bare. When it comes to inferring your personality traits, political views, or consumer tastes, it is close to the mark. You feel panic. Why? Because knowledge is power. If they know all about you and you know nothing about them, you are at a disadvantage in any confrontation or negotiation.[42] They know what and who you resonate with, which lines of argument appeal to you, what you like and how much, what you are willing to pay, when you are bluffing and when you playing straight, your vulnerabilities and addictions. Taken

---

[41] Consider the reaction if the October 2002 Washington-area sniper had been in the pay of hostile states or al Qaeda.

[42] Andrew Odlyzko, an economist at the University of Minnesota, argues that companies that amass a large amount of information on potential customers use this information to charge each the highest price (one that would vary from person to person) that they would be willing to pay for something. Andrew Odlyzko, "They're Watching You," *Economist*, October 16, 2003, p. 77.

to extremes, such a profile leaves you virtually powerless – able to assert power only by withdrawing.

Can warriors in cyberspace, intent on changing national policy or the national will, use personal information to better persuade a country's citizens one by one?

Start with an easier question: what can information warriors do if they focus on learning about influential people? The problem is that the art of persuading already-influential people gains little momentum from knowing their personal habits. Many are under a continual spotlight any-how when in public and they hardly have the time to engage in random unrecorded chat room sessions with strangers; there may be little private information to exploit. They are anyway usually direct about their *osten-sible* views on this or that public issue. Much of what must be inferred about everyone else can be read directly from the public record for such people. Finally, if they are public officials, they may be constrained to stay close to the constituencies, institutions, or movements behind them. Many have taken public positions that they cannot easily abandon even if someone were able to convince them to do so. Information warriors who possess nothing but a little inside information must compete for influence with professional lobbyists with all their tools of persuasion. So, getting to them is not everything; in some ways, convincing the public is more cost-effective.

The use of private information to persuade a public must presume that the information going to each person is individually tailored. Mass mail-ers have found out some of how to do that. Their art is to cull names from lists and use the source of the lists (Second Amendment fans, members of engineering societies, pro-choice advocates, and so on) to concoct a specific pitch to these people in order to support particular causes (such as the election of a friendly representative). They are necessarily one-dimensional arguments, moving from attribute to pitch.

With the ability to collect considerably more information on people plus the ability to individualize messages in cyberspace compared to mail space, tailoring may approach perfection – especially if interaction with the recipient can be used to tune the message. A person who follows NASCAR, groans under a large mortgage, and frequently weighs in on the topic of Armenian genocide may get a different message than someone more interested in expensive suits, obscure novelists, and the impact of Christopher Columbus on Native Americans. Interactions with every individual may evolve in unique directions as each reacts to this or that point made. Exactly how these ancillary attributes can be leveraged in persuasion is a mystery – but American political consultants are likely to crack it as soon as anyone else does.

Such selling points have to be delivered through individual media such as e-mail, instant messaging, chat rooms, personalized Web pages, or, perhaps some day, one-to-one television. Getting someone started on the message is not trivial but not impossible. Today's spam, for instance, is generic and can be recognized by people as such; knowing the recipient permits spammers to write header lines of specific interest, tricking people into reading them – at least until they recognize such tricks. What of other media? Instant messaging and hanging out on MySpace.com tends to be popular among those with far more free time than political influence (such as teenagers) and requires breaking into a group in which all the members may know each other. The Web page, for its part, has to be found in order to be read. One could pay for exposure, but click-through rates on banner advertisements are still low these days. Pop-over and pop-under ads tend to put people off and can be filtered out by, among others, Google's Web taskbar.

Two exotic methods merit mention. One is to hijack oft-visited Web pages so that what the user sees is not only specific to the user (as is becoming more common) but carries the message of the persuader. A similar trick is to put a shunt into networks to divert Web page requests to another site; several pornography and phishing sites try this. Such sites may look like the real thing but with altered or added news or commentary. The trick must be subtle and used sparingly; exposure can ruin the message in all other media. One could use "push" techniques – imagine a park bench that can scan those who sit on it, figure out who they are, and address advertising to them because they know their interests.[43]

---

[43] Steven Spielberg did in his movie *Minority Report*. The cover article of *Wired*'s March 1997 edition bid us all to "Kiss your browser goodbye!" It supplied an interesting vignette of what was called "push" technology, but as with many such articles, it was a tad early.

8.8 Some Limits of Retail Warfare in Cyberspace

Mere knowledge of someone does not make persuading him or her a cinch. Not only do people rarely hold invariant views on a wide range of issues, but they can be opaque to those who know them best: relatives, wives, even themselves. Many of the nonverbal cues that, when read correctly, afford so much information on what people will buy, in both senses of the word, are simply unavailable through cyberspace or even incidental surveillance.

It may become harder to collect information on people if the consequences of a promiscuous sharing of personal information become widely known. A systematic, devious, and, to sharpen the point, foreign exploiter of personal information is bound to pose a larger threat than pesky salespeople who seem to know us a little too well. So, whatever leverage can be gained from personal information can be used only once in a big way, or must be wielded under the radar, so to speak.

Once people find out that they have been spied on for the purpose of political persuasion, they may be reluctant to reveal anything further to anyone about their likes and dislikes.[44] What damage would this do to the persuader's ability to exploit what already exists? What people like, fear, or resonate with does not change greatly from one year to the next, even if their reactions to fashions or the particulars may change on a week-to-week basis. Thus, while some of the value of personal profiles may decrease sharply once data are no longer collected, the bulk of it retains its value for years, especially for those already adult when circumspection set in. But what works for the small (such as one-to-one marketing) may not work so well for the large. The difficulty of predicting public opinion on tectonic issues limits any one-to-one information campaign (especially for a U.S. audience). Many were surprised that the U.S. public was ready to wage the Cold War in 1946 and 1947 in spite of contrary indications, such as the widespread "return to normalcy" sentiment following World War I.[45] The

---

[44] So far, however, no such violent reaction has followed the news that we are being spied on for purposes of market persuasion.

[45] Eric Goldman, *The Crucial Decade and After: America, 1945–1960*, New York (Random House), 1972.

outpouring of "United We Stand" pronouncements with their implied inclusiveness after the September 11 attacks was hardly obvious in their immediate aftermath. Many elections turn on last-minute reactions to events that took place weeks before but had yet to register fully with the electorate.

Any retail campaign also faces competition from the constant drumbeat of information from wholesale sources. Even today, news sources and influential opinion makers speak to a global audience.[46] News events will have similar aspects regardless of who views them. A government's press announcement is the same to all. The main lines of the argument are likely to be global even if people focus on one aspect or another of a news story. The truth cannot be exquisitely tailored to every individual. Even if the line spreads solely through cyberspace – whether pulled from individualized postings or pushed via e-mail or IM – people will talk to each other. Inconsistencies will out. There must be some line to be peddled that is common to all.

Millions of profiles are, anyway, overkill if someone wants to determine what to pitch – a statistically valid sample of a few thousand will suffice. More broadly, personal information, especially of past commercial and/or political transactions, is far less important in selling a message than is a shrewd understanding of the audience.

8.9 Using Retail Channels to Measure Wholesale Campaigns

Nothing in this chapter suggests that wholesale messages, having lost their exclusive place in the information warfare pantheon, must therefore disappear. Every nation or faction is expected to account for its actions, explain their rationale, and put them in the best possible light. Every White House must have its message of the day. But then there's that tricky issue of battle damage assessment: is the message having any effect? Hence, there is another use of retail information warfare: to find out how well wholesale messages are resonating.

---

[46] Following what to many was the unexpected reelection victory of President Bush, several commentators argued that the Republicans had found a way to communicate with religious voters in ways that were not picked up by the national media.

Part of the answer has to do with the structure of the message itself. To start, let's review some basics. People tend to understand the world through narratives, a selection of tales that collectively form a worldview in the sense of a global menu of moral possibilities. Each culture favors its select narratives. Schools teach some of them explicitly, and they are socially reinforced within communities. Not all narratives are universal; many cultures dwell upon their own victimization by the rest of the world, an attitude that, by its nature, has to be peculiar to them alone.

Narratives may persist even if the decisions they inform seem to shift sharply. The terrorist who gets married and leaves the business may not necessarily do so because he is convinced that terror is evil or that the cause for which he fought is no longer valid. Perhaps the terrorist, who formerly saw his identity as that of being a vigorous defender, has switched the locus of this defense from his "beleaguered" people to his immediate family and then realized that the family cannot be defended if he is gone. Here, the basic narrative that gives his life meaning does not change, just the character roles. The case of committed Communists who became committed anti-Communists may not have been so distant a switch if both perspectives were grounded in the same narrative: both believe that history is made by conspirators that they have a duty to expose.

At a minimum, psychological operators have to understand the narratives current in the target population, because that is the lens through which the news is viewed. Then the trick is to move people away from unhelpful narratives, but not necessarily by battling the narratives themselves. Most have a basis in reality – after all, there *were* Crusades, a long history of Christian–Moslem rivalry, and colonialism in the Islamic world; convincing Moslems that Christians were the good guys in the Crusades is largely a waste of effort. Instead they should try to do two things: (1) argue that the narrative does not apply in the circumstances ("neocolonialists, us?"), and (2) bolster a substitute narrative in its place that assigns people correctly to their roles ("democracy good; Zarqawi, being antidemocratic, bad").

The relationship between narratives and vocabulary is fairly straightforward. A narrative represents a tacit generalization about the world. Every generalization, for its part, is an attempt to organize a mass of instances around specific characteristics. A person may be Sunni or Shia, pro-democracy or pro-autocracy. Someone who believes that ethnicity is all (that is, that all relationships can be viewed within the context of the power struggle between Sunni and Shia) is likely to use use ethnic characteristics to make generalizations; someone who believes in universal human values is far less likely to do so. Words, for their part, also reflect generalizations, particularly as they become more abstract. Hence the correlation between words and narratives.

A semantic analysis is thus a strong clue as to which narrative is prevailing.

One can then use the techniques of Chapter 5, ping and echo, to see how well messages are playing. Here, the wholesale message, the psychological campaign, is the ping, and the echo is the response to the campaign. The more psychological warriors are wired into the other side, via taps into the target's cyberspace, the more sensitively echoes can be detected. One then looks for which narrative is playing most heavily. Sometimes this information will be volunteered in one-way conversation (for example, blogs and their responses) and sometimes in two-way conversation (IMs, chat forums, and such). There will also be times when the conversations do not come up at all; people have their minds elsewhere. This is when to use some techniques discussed in this chapter, notably techniques of simulating conversation and collecting responses to it.

A wholesale campaign can be measured statistically and thus fairly accurate readings can be obtained by sampling. By contrast, a retail campaign to convince everyone, a person at a time, has to measure each engaged individual as part of the process by which the argument is brought to bear. The latter measurement sounds more difficult, and it usually is. Yet, measuring the response to a wholesale message may elicit what people thought before being engaged on a one-to-one dialogue because those being sampled are only a proxy for a population that, by and large, is *not* engaged in such one-to-one dialogue. Thus, too intensive a dialogue for testing the message may distort the readings. Getting a response to a wholesale message by asking about it in conversation does not mean that the wholesale message delivered in wholesale channels is getting such a response or, indeed, any response.

8.10 Conclusions

The promiscuous proliferation of personal information via cyberspace may tempt tomorrow's information warrior to exploit such information for writing one-to-one persuasion campaigns. The challenge they face in amassing this information, not to mention exploiting it for strategic effect, should not be underestimated. Yet, if finding the right set of techniques coupled with an effective overall campaign spells success, exploitation may only be a matter of time, and not much time at that.

This prospect suggests a complex interplay between friendly and hostile conquest in cyberspace. Much, perhaps most, information collected on people is amassed with their consent (although exactly how informed they are is another issue) or at least their acquiescence. Strangers can learn a lot about people in real space through charm and cleverness – and charm and cleverness could become be an attribute of bots one day. Vendors such as banks learn a lot simply through continual interaction. Getting such information from the right hands to the wrong hands is where hostile use of cyberspace comes in handy.

Greater security in cyberspace may enhance privacy if it prevents third parties from stealing information collected by second parties; indeed, protecting client confidentiality may well be the primary reason that corporations take security seriously.[47] But where computers and their uses must identify themselves to interact, security and privacy can be antithetical. The security and addressing features of IPv6 allow users to associate addresses with persons permanently, making it easy to correlate virtual and real identities. Even without IPv6, if cyberspace were to become more dangerous, digital signatures may become a prerequisite to accessing many Web services. Digital signatures characterize people rather than machines; again, the virtual–real correlation is bolstered. All

---

[47] Although enterprises whose security has been violated may want to cover up that fact and everyone inside may cooperate, those whose privacy has been violated have no such interest in hiding what happened. There is no evidence of any absolute upper limit to the damages some court could levy as the penalty for allowing such leaks to take place. Nevertheless, as a general rule, privacy protection may be the lever that makes administrators serious about defending cyberspace. See Alice Lipowicz, "Privacy Matters," *Washington Technology*, 20, 10, May 23, 2005, pp. 1, 30–2.

this facilitates the construction of personal databases for exploitation by information warriors.

It remains unclear how much the exploitation of personal information will be worth to psychological warriors. But the more material piles up, the greater the potential rewards of finding an innovative way to combine friendly and hostile conquest. Who can say that someone will not succeed in turning mostly commercial dross into propaganda gold?

9

From Intimacy, Vulnerability

Close relationships in cyberspace, as in real life, can make either partner more vulnerable.[1] A relationship, solely by virtue of the value it brings to its partners, may be attacked by competitors of each. Third parties can exploit weaknesses in one to get at others. More insidiously but, fortunately, less commonly, one partner may exploit the relationship to wend its way into – and from that vantage point, to assault – the systems of partners.

Love and war, so famously mated in fiction, find echo in cyberspace. Here one asks: by what means, and with what effect?

9.1 Do the Walls Really Come Down?

Two organizations that would exchange information on a continual basis can most seamlessly do so by merging their respective systems, thereby creating a common cyberspace so that members of one gain full access to the resources, data, and applications of the other.

---

[1] As noted in an *Economist* survey:

All these threats arise from a common factor: the distinction between the "public" parts of a company's network (such as the web servers where its home page resides) and private core (which is accessible only to employees) is quickly eroding. "The cultural and technological trend is toward more porous companies," say Gene Hodges, president of Network Associates, a large security software firm. As firms connect with their suppliers and customers, "the more you open up the more you are exposed" ("Securing the Cloud," *Economist*, October 26, 2002, p. 17).

Yet, with all the risks of intimacy[2] that arise from unprotected networking, are enterprises so likely to lower their barriers entirely? Furthermore, is it that necessary? After all, internal barriers are routine even within single organizations. Rightly or wrongly, many organizations maintain a hierarchy of privilege with respect to seeing information. Others adopt compartmentation in the belief that any sufficiently interesting piece of information known to enough people within an organization is bound to be known to the wrong person outside of it. Similarly, any corruptible systems process accessible to enough people on the inside will eventually become corrupted. Even those without classified work have information they want to show only to trustworthy outsiders and they will go through considerable trouble to determine who can, in fact, be trusted.

Can the right controls permit intimacy without risk? In some cases, one can replicate internal access controls to cover outsiders much as DoD extends privileges to defense contractors with security clearances. Information can be categorized by who is allowed to see it.[3] What can be shared with external individuals would be placed outside the enterprise in a virtual place to which both have access. Such controls can be specified with any requisite precision, such as "he can, you can't."

---

[2] Joint ventures, in which two firms collaborate and share information, can also cause problems. A recent report by Vista Research cites the example of an American car maker that established a joint venture with a Japanese firm and opened up its network to allow in employees of its Japanese partner. But the design of the American firm's network allowed access only on an "all or nothing" basis, so the Japanese firm's employees ended up with access to everything. Ibid.

[3] A big problem is limiting transfers of information that comes from one party to another party and from there to unauthorized third parties. Unless the relevant information is compact enough (such as news of an impending merger) to be transferred by hand or mouth, such transfers must be digitized. Digital retransfer could be impeded by limiting read operations to terminals without transfer mechanisms (for example, terminals with no removable storage or network links), but this is generally impractical. Information could be encoded so that it may be read only by applications designed to limit transfers. The digital content industry (producers of music, E-books, and so on) is trying hard to make such an approach work. Unfortunately, once such applications are made public, some hacker tends to break them. People are also working on methods to watermark information so that who was entrusted with the information that leaked can be determined by intercepting the once-too-often-copied information. Yet, while pictures and music watermark well, they too can be broken; text and databases do not offer enough bytes to watermark uniquely.

Those worried that intimacy can lead to the backwash of virtual vermin could pass information back and forth to a separate facility using dedicated communications paths and with protocols that permit nothing but data to be exchanged.

In either case, peace of mind comes with prerequisites. One must know exactly which information is to be transferred into and out of the shared facility, as well as the procedures for synchronizing and updating information within the facility to match its counterparts behind the walls. Transfer mechanisms must come with all the requisite fail-safe, roll-back, and graceful-degradation features considered standard, but not always sufficient, in modern software.

The rest of us must use common software systems and protocols such as Windows, Linux, and TCP/IP. All have features that are growing continually more complex, harder to manage, and more tightly coupled with applications software. Each makes not-always-obvious use of not-always-apparent function calls with their not-always-intuitive models of safe and unsafe access practices.

And that's today. Coming soon are applications that scan the globe for knowledge (for example, they might monitor medical records to build a knowledge base of diagnoses, protocols, and prognoses) or negotiate for goods and services on behalf of their owners. Whenever such software can only guess the protocols of its collaborators, it must then probe them to ascertain whether its assumptions about them are valid. Without a common set of protocols, there is no way to know for sure what comprises a complete combination of states, calls, and responses.

The problems of intimacy cannot be engineered away so easily.

9.2 Intimacy as a Target

Third parties may want to attack a relationship simply because it is the heart of an opposing alliance. Sundering relationships can render opposing alliances less effective.

The logic of breaking mirrors the logic of binding. The ability to form coalitions, as argued in Chapter 6, is of growing value in competitive arenas. Much as personal relationships are ratified and maintained through the exchange of favors, coalitions float on the exchange of information – especially the privileged exchange of sensitive information such as inventory data (as in Proctor & Gamble's evolving relationship with Wal-Mart), design information (as for new cars), or customer profiles. The greater the importance of proprietary and personal information flowing among enterprises, the more important to alliance cohesion is the ability and willingness to protect such information. Thus the more important good security is to the choice of partners. Continuous up-time and information integrity are a must. Having to repudiate false messages in today's hyperspeed financial markets, for instance, could force partners to retract several layers of exchange with outsiders. Information systems, if successfully attacked, lose their credibility as a basis for alliances as well as their overall efficiency.

An attacker who soils someone's reputation as a partner also lowers what the victim can ask for as a price of its membership in any alliance, permitting the attacker to offer future partnerships with the victim at cut-rate prices.[4] With competition for partnerships so intense, any reduction of a competitor's reputation for trustworthiness in cyberspace redounds to another's benefit. Even though the other bankers unanimously condemned the Russian hacker who looted Citibank accounts for up to $10 million (albeit 95 percent was recovered[5]), they were not shy about casting aspersions on Citibank's ability to protect itself – and, by extension, its customers.[6]

For those who do not see economic competition among nations as a zero-sum game, queering business coalition formation may be one of the few valid nonmilitary motivations[7] for government-led computer

---

[4] To conclude that its victims are not inherently more vulnerable than their peers, the attacker must believe it has unique ways of getting into systems that are not characteristic of other hackers, or that everyone is vulnerable to that particular attack but only the victim's vulnerabilities have been demonstrated; those organizations considering membership with the victim do not realize (or serve others who do not realize) that the victim has been singled out. Otherwise, by probing the victim and finding vulnerabilities, the attacker should learn that the victim's insecurity makes it an unworthy partner and that the low price is, in fact, merited.

[5] "Could a Cyberterrorist Take Down Your Company?" Knowledge@Wharton (2009-12-956901), September 8, 2002.

[6] See "Cyber Crime," *Business Week*, February 21, 2000, p. 41.

[7] Direct attack on a competing nation has little to recommend it except as a part of a broader campaign to disarm foes in some decisive and persistent matter – which, of

network attack. The attacker need not be the potential alliance partner itself. States, eager to see their corporations strike useful deals in the global marketplace[8] and self-righteous enough to protect their hackers may imagine themselves to have enough reason to bias strategic business relationships in their companies' favor. A state that sponsored such mischief could even condemn the result, support the victim, pledge cooperation with the investigators, and perhaps even turn over some of its internal traffic – as long as it is careful not to incriminate itself. Alternatively, it could take a harsher approach: regretting inconvenience to the innocent, wondering whether the accused had not itself been the victim of past misdeeds, condemning the victims for casual security practices and maybe even having provoked the ire of the attackers, and closing itself off to any investigation that would prove otherwise. A hard attitude may be justified on the grounds of state sovereignty as well as distaste for the victim's government (the entity that would be carrying out the investigation) coupled with the inevitable claims that such an investigation was a cover for other forms of state mischief.[9] This posture fails only if enough of the world holds its norms of a noise-free cyberspace environment so strongly that it would condemn any state that holds these norms with insufficient reverence or shows itself to be opaque where transparency is called for. Even if caught, the attacker's sponsors may gain if potential coalition partners remain more averse to linking their systems with the hapless than with bullies. They may rationalize that everyone would like to be nasty, but some are more successful than others.

Caution is, nevertheless, called for when invoking this scenario. In today's business environment, competitors at one level may want to

---

course, raises the specter of real fighting. After all, a country could have to become quite entangled in theologically pointless distinctions to declare war on the United States in one medium, confident it would not be counterattacked in another. Attacks that do not rise to the strategic level often do little but create enemies through mindless destruction (or at least mindless annoyance). In any case, the coercive potential is lost if perpetrators cannot identify themselves.

[8] This also presumes a relationship between states and "their" corporations, which seems dated in the West, but retains relevance in Asia. In some countries, government and business are underwritten by the same criminal organizations.

[9] Investigations into hacker attacks on an unclassified Pentagon network in 1999 (Moonlight Maze) were stopped cold when a Russian Internet service provider refused to cooperate.

cooperate at another, and alienating rivals leads to fewer opportunities for profitable cooperation. If cyberspace becomes more seriously dangerous, coalition partners in the West may also vet less-established applicants more strenuously. In effect, they would raise the technical requirements (such as computer security) for membership. Both would raise bars to the less well-established companies likely to be among the attacker's friends.

9.3 The Fecklessness of Friends

Even in a world without nasty competitors, any organization that has tied the health of its cyberspace to the security consciousness of its partners has cause to worry. Sure, *all* dependency relationships, even those that do not exchange much information, are causes for concern when hostile conquest is afoot; everyone shivers if the gas company's services are disabled by hackers. But an organization that exchanges information – especially if it creates entrée for embedded instructions or requires granting partners specific privileges – has more to worry about. It can attend to its own security and maintain a good sense of the trustworthiness of its partners – and justifiably look askance on opening its cyberspace to others.

How can coalition partners assess the susceptibility of each other to attack? How can they use this information to understand how much privileged access to accord them? Few organizations will confess to their own blithe sloppiness when it comes to information security. Only a few will turn over logs of specific failures to their partners – and the value of such logs assumes they could be meaningfully mined by those without detailed knowledge of the system that was compromised.

A preliminary step in vetting potential partners is to think through whether the intimate combination of two enterprise architectures results in an agglomeration less secure than either. This can happen, for instance, if two systems that allow easy access to one another put their protection shells in different layers. System A, for instance, may ensure that users cannot assume super-user status but take fewer pains keeping outsiders from becoming legitimate users. System B may place its perimeter further out, erecting barriers against outsiders but giving insiders freer play.

Thus a hacker may hurdle the relatively low defenses of System A, and once inside, use the partnership between A and B to assume privileges on System B. The hacker then exploits the greater license that System B gives to users to wreak havoc.

Several methods can be used to assess the diligence of partners.

One way is to view a partner's system as a black box, see to it that its defenses are attacked, and observe what happens next. Unless one wants to test the system against insider mischief, the hired attackers should start without any information gleaned through privileged exchanges. Nevertheless, even if one has permission to do so, this *is* a hostile method and thus a poor way to maintain a relationship supposedly founded on trust. Not all results will be externally obvious: the damage may be deep or the victim may have work-arounds that effectively mask damage done to the control system.

One can also read the partner's documents: intrusion logs, software configurations, the policies by which it regulates access, the frequency and care with which it applies software patches, and so on. This admittedly labor-intensive method is useful, but provides only a snapshot.[10] The next installed software or sloppily installed firewall can create new pathways for the mischievous. A census also does not take the more dynamic aspects of systems protection into account. Installing an intrusion detection system is an act that, once done, can be checked off. Knowing that a partner responds to intrusions in a useful manner is a different matter. Without detailed knowledge of the system's architecture, it is difficult to know the effect of an attack on the partner's system itself – much less

---

[10] How reliable is process evaluation for security? The NSA ratings of systems (see *DoD Trusted Computer System Evaluation Criteria,* December 26, 1985, or other reports listed on www.radium.ncsc.mil/ttpep/library/rainbow) are primarily meant as judgments of software (and hardware with software attributes such as chips with embedded logic). Their purpose is to guide assessments of whether such software has been created in such a way that the good faith of their creators can be trusted. Companies usually do not worry about the latter, and even knowing the software is good does not prevent such security holes from accumulating due to poor installation, failure to maintain security patches, or faults outside the software (for example, control systems that do not keep hardware hazards from resulting from software faults). The criteria of the more recently developed Common Criteria are also difficult to audit unambiguously, although NIST and NSA jointly manage the National Information Assurance Partnership Common Criteria Evaluation and Validation Scheme (CCEVS) Validation Body. The newly touted Generally Accepted Security Principles are yet harder to audit against its criteria.

its ability to compromise one's own systems. Knowing how assiduous a partner's systems administrators are may say more about their security than knowing exactly which patch sequence they are working on.

Systems security can thus be assessed not as a state but as a *process*. The ISO-17799 standard[11] could do for systems security what ISO 9000 did for quality control and ISO 14000 for environmental management. Quality-control and environment assessment has drifted from outcome- to process-based evaluation because measuring either product quality or environmental control requires detailed expertise. The quality of an electronic component or a machined gear is hardly obvious upon inspection; determining as much by looking at the overall process requires expertise only in the replicable features of its overall management. The same holds for environmental protection: for chemicals, traceability is important; for power plants, it is emissions control; for biotechnology, it is release testing. One process-auditing mechanism can cover a heterogeneity of production types. Separately trained audit staffs are not needed for each area of concern. Management practices can be compared on the theory that an organization with a verifiably sound method for avoiding certain types of errors (such as defects and pollution, respectively) will make fewer of them. Security follows similar rules; one need not be a security expert to gain or lose assurance in the quality of a process.

That noted, the information systems of different sectors might be more similar than each sector's quality-control or environmental management requirements. Some software (including Microsoft's and Cisco's) is the same everywhere – so, similar actions are required everywhere. Office automation, Web hosting, and database management have many similar elements regardless of who runs them. At the risk of oversimplification, whatever security faults are associated with the implementation of Microsoft Vista are unlikely to vary much from one industry to another. A great deal of security monitoring is a variation on a theme: can outsiders be kept out; can insiders be kept from doing what they ought not; can systems gone bad be expeditiously recovered; and is hardware sufficiently protected from software faults? So there is far less need to look at process as an always-approximate proxy for a product because the products have a great deal in common.

---

[11] ISO-17799 is the internationalized version of what was originally British Standard 7799.

Suppose one partner *did* take upon itself the role of preventing attacks on all the others. The motive may be quasi-altruistic; for example, a government takes on this role in order to protect the nation's infrastructure, or a prime contractor expresses a similar concern for its suppliers' systems. This can be done with minimal knowledge of partners' systems (letting them know of virus alerts or bug fixes, for instance), modest knowledge of partners' systems (by using information about attacks on one to warn others, for example), or deep knowledge of partners' systems. The latter can range from sending so-called tiger teams on "request" to bolster the security of partners or help them recover from attack. Or, it can include an active role in network management. The latter requires considerable trust in the managers sent over to do this – not only in their honesty but responsiveness: will external managers restore service to their clients with the alacrity that they would have restored their own service? In any case, there are companies whose business *is* to protect other people's networks for a living.

Here, as elsewhere, letting others do the hard work over time makes it all the harder to tend to such tasks oneself. Just as Rome's neighbors found that it was easier to ask Roman armies to protect them than to get the Romans, having finished up, to leave, letting others manage one's security may make it all the more difficult to leave the partnership later on. A sneaky partner may have to move matters along by persuading its lazy colleagues that only a nonstandardized security management approach will do. With that, the friend embeds its own "monitoring" programs into its partners' systems that are difficult to find, too intertwined with the watched systems to remove cleanly, and if, left in too long – well, idle software agents are the devil's workshop. This dependency, needless to add, is not specific to "security" monitoring; any support relationship carries the risks of leave-behinds to act as a back door to be entered for old times' sake.

9.4 Betrayal

What should one make of the possibility that one partner in an alliance will use privileged access to another's system to conduct information warfare on it? Much depends on whether friendly or hostile conquest is at issue.

Someone granted privileges to a partner's system may be able to troll around in it (to learn how it uses terms and tag sets, for example) in order to sprinkle "lubricant," such as pieces of code, object methods, tag definitions, or engineering rules of thumb. This can reduce friction and thereby increase the efficiency of joint and combined operations. Because the effort may increase the difficulty of leaving, even as it offers short-term efficiency gains for those who stay, the offer is one that may be rejected – if it is known about. So, here anonymous altruism is a betrayal of trust – albeit one that remains mostly hypothetical.

The more normal forms of mischief are to use privileged access to induce the three faults from information warfare: corruption, disruption, and exploitation. Exploitation is clearly the most plausible and the least likely to result in the perpetrator getting caught, because little of great obviousness is changed in the target system. Many types of information merit theft, including customer lists, negotiation strategies as revealed in e-mails, research results, clues to potential hires, and software code (although a byte-for-byte copy can be legally implicating if revealed). But would the attacker need to take great risks to steal that which a close partnership will often give it through day-to-day interaction? The normal course of business suffices, in fact, to suggest to each one partner which employee is worth hiring away, what their negotiation or business strategies are, where their technology is going, and what future products they are hoping to introduce. Microsoft, for one, has picked up a reputation – however unfairly or not – of making partnerships with smaller firms, asking the sort of questions about their partner's software that are consistent with building hooks into their own software, abandoning the relationship, and coming out with an all-too-similar feature or product within a year.[12] None of this requires hacking into the systems of others.

---

[12] Sendo's suit against Microsoft is only one recent example. Matthew Broersma, "Sendo Accuses Microsoft of Dirty Tricks," http://news.com.com/2100-1033-978687.html, December 24, 2002. A *Business Week* article has noted:

> Several companies charge that Microsoft has, in effect, stolen their ideas in the course of collaborative agreements. Go Corp., for example, says that Microsoft expressed interest for Go's OS for pen-based computers. After Microsoft programmers examined Go's technology, however, Microsoft said it was no longer interested, Go says. Then, Microsoft announced plans for a competing system, developed in part, by those who visited Go ("Is Microsoft Too Powerful?" *Business Week*, March 1, 1993, p. 33).

It is harder to see why business partners would wish to disrupt or corrupt the information systems of one another. On a day-to-day basis, they have (or at least purport to have) a stake in the smooth and trustworthy functioning of their partners. If the attacker can create such problems and determine that such problems have been created, could it recommend itself as the solution – thereby either selling its services, or, worse, worming its way deeper into its partner's systems? A sporty move this, and one which cannot help but stoke the suspicions of its partner. If two companies link up and one is dogged by mysterious attacks, the other is likely to be the first accused and has at least some incentive to demonstrate otherwise by word and deed.

9.5 Conclusions

As President Warren Harding bemoaned after the Teapot Dome scandal went public, "I can take care of my enemies but God help me from my friends."

Friendly and hostile conquest in cyberspace oft intertwine; friendly conquest binds partners closer and, in so doing, makes them more vulnerable to hostile conquest. Relationships carry with them vulnerabilities, from the potential malevolence as well as the sloppiness of one's partners. But the worst possibility may be abjuring such relationships altogether, something that global rivals may want to see happen more often. In any case, attention to the terms of future alliances as reflected in cyberspace is necessary.

For, in cyberspace, good fences *do* make good neighbors.

10

Talking Conquest in Cyberspace

Early computer systems were machines used as tools to perform given tasks in repeatable and predictable ways. Networking, as such, did not kill that metaphor. A distributed system is one with its components here and there; inputs may come from hither and outputs may go to yon. But the system retained its machinelike functionality, structure, and process.

Cyberspace, though, is harder to liken to a machine. Its components are not only distributed but also often independently controlled, especially when they interact. At first, cyberspace was a medium through which people alone exchanged information, a phone system as it were but one where the content that people traded – text, image, audio, video – was all turned into bits. But people also interacted with the system and its artifacts, such as files or programs. Increasingly, processes (such as shopping bots), nodes, servers, computer-controlled machinery, and so on, entered cyberspace to exchange information with each other.

Cyberspace is thus less of a passive medium and more like a stage that both supports and interacts with billions of entities, humans among them, which themselves pursue billions of purposes in conjunction with one another, with all assertions, comparisons, judgments, challenges, and the presentation of bona fides[1] one expects in conversation.

---

[1] Reports Barry Brigs, "WS-Security details the mechanisms by which a SOAP [Simple Object Access Protocol] message can describe not only its contents but also its credentials and the algorithms by which your XML packet's privacy and nonrepudiability were generated." Barry Brigs, *Infoworld*, June 17, 2001, p. 58.

What does it say about conquest in cyberspace if systems loss their machinelike character and cyberspace is seen as an enormous conversation? Examining analogies between human linguistics and emerging computer processes, this chapter explores:

- How confusion in the syntax-semantics layer facilitates conquest – both hostile and friendly
- How the semantic layer is becoming a vector of friendly conquest
- How the pragmatics layer may one day affect hostile and friendly conquest.

10.1 Four Layers of Communications

Although the layering of human conversation has direct and suggestive analogy with the layering of computer interactions, interoperability conventions play a much greater role in the latter. People, being smarter and more adaptable than machines, do not need to follow convention strictly to make themselves understood.

10.1.1 Human Conversation in Layers

Human language can be understood in terms of four layers: phonology, syntax, semantics, and pragmatics.

Phonology deals with how people manufacture sounds by using various parts of the mouth and larynx. Speech is prologue to writing; almost all languages were first spoken and their grammars still bear the limitations of their vocalized provenance.

Syntax covers grammar, from the morphology of word construction (suffixes, for example) to word relationships and word order. In English, syntax is critical to making subject/object distinctions, establishing chronological order, and adding redundancy to error-prone communications.

Semantics is what words mean. Categorization is a key component of syntax in that words are often defined as sets, that is, in terms of what they include. For instance, some languages lump green and blue together as one word; English separates them. Categories, in turn, are the basis for generalizations (for example, most A are B) and rules (if x is a C, then do D, for example).

Pragmatics is the implied purpose of a statement or speech act – what it is trying to achieve – in a specific context. Telling a child, for instance, that she did not brush her teeth is not meant to convey information. She, herself, knows that her teeth are not brushed. The statement is, in effect, a command.

In general, the ability to separate a complex set of relationships into independent layers can help understanding. One need worry only about each layer at a time. But is strict layering a useful way to examine language? Were it so, then how words sound would not necessarily dictate or even influence how they should be sequenced in a sentence; syntax ought to provide a skeleton into which one can pour all sorts of words; different statements can, in turn, serve analogous pragmatic purposes (such as conveying information). In truth, the four layers *are* entangled. Many irregular verb conjugations, at least in English, arose because of shifts in pronunciation over time.[2] Many words are onomatopoeic; the meaning of "squish" is suggested by its sound. The semantics of a simple word such as "there" is altogether related to its syntax; for example, "Now, there is a man," does not mean exactly the same as "a man is there now." The pragmatics of a sentence in turn may be affected by what is labeled mere semantics: "Are we going to *walk* to the mall?" is a question; "Are we going to *trudge* to the mall?" is a complaint.

Interdependence, in practice, is no bar to speech or writing because language is learned as an integrated whole. People use it intuitively. Although phonology tends to be learned earlier in life than syntax, syntax earlier than semantics, and semantics earlier than pragmatics, children do not wait to master one layer before embarking on the next. They pick up some pragmatics even while learning how to sound words (phonology).

By contrast, the requirement for layer independence is stronger in cyberspace because computers are simultaneously more complex and brittle. They balk, freeze, carry out orders that would violate common sense if given to humans, and lack the sort of somatic indicators (for

---

[2] This is pointed out in Steven Pinker, *Words and Rules*, New York (Perseus), 1999.

example, fear of falling, pain avoidance, and physical attraction) that inform human action even in the absence of language.[3]

Semantics is where human and computer layers begin to interact on something of an equivalent basis (by contrast, human syntax has very little to do with computer syntax, except metaphorically). Words are the vessel through which knowledge is transmitted because they structure categories, the basis for generalizations and rules. Suppose we teach children never to cross highways; cross streets only after looking both ways; but walk freely everywhere else (as when traversing trails and sidewalks). This lesson presupposes some common and common-sense notion of what constitutes a street. Do they include back alleys? Must they be paved? Do dead-end streets count? How does one differentiate between a street and a highway – what volume of traffic or access controls make a difference? True, one could make fine distinctions by taking gross categories and qualifying them (for example, comparing streets that have outlets to dead-end streets), but practice tends to generalize simple categories. We lack the experience or statistics to create rules for microcategories, and people lack the attention to remember each of these microcategories when they apply such rules anyway.

Where do such categories come from? The objectivist school is Platonic; it argues that categories are representations, with greater or lesser fidelity, of inherent distinctions among objects or actions. The very essence of a chair and the essence of a sofa differ greatly; that's why different terms are used for each. George Lakoff argues that no such external objectivist standard exists.[4] People group or differentiate things based on how they relate to them, and typically, with their bodies. Chairs, as such, are not ideal subtypes of furniture, but things that we relate to by sitting on them individually; sofas may look like chairs, but we can lie on them or share them with others. Furthermore, "chair" is a category that is more basic than the superset "furniture" or the subset of chair, "recliner."

---

[3] Hubert Dreyfus argues that because computers do not have bodies (and thus do not know how things relate to bodies), they cannot have the common sense that people have. Herbert Dreyfus, *What Computers Still Cannot Do: A Critique of Artificial Reason*, Cambridge (MIT Press), 1994.

[4] George Lakoff, *Women, Fire, and Dangerous Things*, Chicago (University of Chicago Press), 1987.

Boundaries between colors stem from the electromagnetic frequencies that eyes were built to distinguish. Stephen J. Gould, noting three fundamental types of zebras, one of which was from an entirely different origin, argued against easy criticism of the human tendency to lump them together (for example, categorizing them based on appearance) rather than using objective criteria for differentiation (that is, cladistic criteria, or common evolutionary ancestry).[5] Neither criterion, he argues, is ipso facto superior to the other.

Human experience[6] structures our vocabulary. But human use is not necessarily optimal for making distinctions that lend themselves to good logical and accurate generalizations. (For example, does it make more sense to call a computer anything that computes or something that people use as they would use a personal computer?) Why must the categories that machines use to think with accord to the fundamentals of human experience? After all, a lot about how we live our life has changed since words were coined eons ago; and it keeps changing.

To what extent do the actual words in a language – what actions or things are lumped together as one concept – limit and channel human thought? The Sapir-Whorf hypothesis that they do arose from the discovery that some Amerindian tribes lacked certain words and never uttered statements that used such words – suggesting they were incapable of thinking about the concepts and distinctions that the missing words would have defined.[7] In George Orwell's *1984*, Newspeak was a desiccated version of English with key words, such as "freedom," excised. The lack of any words with which to express subversive thoughts was said to make it difficult or impossible for dissidents even to think them.

Linguists, on the whole, do not find the hard version of the Sapir-Whorf thesis to be well supported. People from such Amerindian tribes could readily express concepts such as mathematics in English once they

---

[5] S. J. Gould, "What, If Anything, Is a Zebra?" in S. J. Gould, *Hen's Teeth and Horse's Toes*, New York (W. W. Norton), pp. 355–65.

[6] Specifically, the experience gained when the distinctions were originally made centuries ago.

[7] See Edward Sapir, *Language*, New York (Harcourt, Brace, and World), 1921; Benjamin L. Whorf, *Language, Thought, and Reality: Selected Writings of Benjamin Lee Whorf*, ed. John B. Carroll, Cambridge (MIT Press), 1956.

had learned them – even if expressing them was more difficult than it could have been. More bluntly, nothing human is untranslatable if one is willing to use enough words to make the same distinctions in both languages. Nevertheless, it is also hard to support the antithesis to the Sapir-Whorf hypothesis – that differences among languages, notably, the categorizations that each one holds, have *no* influence on cognition.[8] One difficulty in proving or disproving the hypothesis is that culture, cognition, and language grow up together. Words evolve to help people talk about what surrounds them. If people in a certain environment need to think deeply or frequently about a topic, the language they use to do so tends to accommodate that desire; if it is not worth thinking about, the words used to ponder them atrophy. Being able to differentiate various types of snow matters to Eskimos (and skiers, for that matter). Patterns of words and habits of thinking coevolve.[9]

As knowledge processing by computers and thus in cyberspace advances, standard terms for real and abstracted entities will necessarily have to precede the rules, generalization, and hence algorithms that exploit these terms. Not only *can* semantics precede cognition, but, in programming terms, it has to (although cognition, in turn, can be the test of semantics) – even if the algorithms that would use such terms are likely to be based on human knowledge built by human cognition and mediated through *human* semantics.

10.1.2 Cyberspace in Layers

Cyberspace can also be discussed in physical (analogous to phonological), syntactic, semantic, and pragmatic terms:

- The physical layer consists of hardware: processors, storage, switches, routers, handsets, and conduits both wired and wireless. It also includes the signals that travel along this hardware.

---

[8] See, for instance, the description of David Gil's work in "Babel's Children," *Economist*, January 8, 2004, pp. 69–70.

[9] See Gün R. Semin, "Language and Social Cognition," in Scott Tindale and Michael A. Hogg, ed., *Blackwell Handbook of Social Psychology*, Oxford (Oxford University Press), 2000, pp. 159–80.

- The syntactic layer contains the programs and conventions by which information is formatted and by which systems (such as computers and networks) are controlled. Syntax governs physical connectivity (for example, the 4/5 framing protocol for fiber-optic networks), correct bit transfer (such as encryption, error correction, and compression), addressing (TCP/IP, for example), quality-of-service parameters, handshaking and session management (as in hypertext transfer), message formatting, database schemas, and markup languages. As it so happens, packets, the fundamental carrier of networked information, use their headers and footers to tell a network how to treat information; the remainder is the information itself. In common with human language, IP networking uses in-band signaling for syntax.[10] Correct headers and footers are critical to the syntactic integrity of information transfer.
- The semantic layer contains the information held and manipulated by systems, plus the transformation rules that manipulate knowledge in high-level applications.
- The pragmatic layer, were it one day to exist, would deal why a statement has been uttered or a message has been sent – that is, its purpose when considered in a particular context.[11]

For two systems to communicate, each layer on one must work together with its counterpart layer on the other, as Figure 7 shows. To be executed correctly, a request must (1) be conveyed, (2) make sense, and (3) use commonly understood terms. To be executed as intended, a common understanding of the intentions and context of the requestor may have to be conveyed.

Many of the standards currently associated with interoperability, especially for data communications, apply to the syntactic layer. Because the layer is so busy, it is generally thought of in terms of its sublayers, of the

---

[10] This feature is not inherent in computing as such. Circuit-switched networks (such as telephones) use a separate channel to pass information on how to handle and route content.

[11] Pragmatics can shed light on whether a request will be accepted. Acceptability is not only based on words of the request themselves but also on the requester's intentions and the context in which the request was made. Often requests cannot be satisfied directly; intention and context suggest when other responses can satisfy the request.

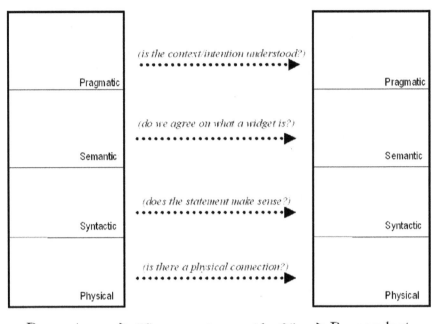

Figure 7. Interoperability at Four Layers

sort defined by the OSI reference model for data communications. OSI was defined in the late 1980s by the world's central *official* standards body, the International Organization for Standardization (ISO). It defines each of seven necessary layers and in ways that made them mutually independent. For two systems to interoperate, each had to exchange each layer's information with its counterpart in a standard way. The reference model thereby defined and limited the scope of specific standards. One could swap out a standard in any given layer with another standard without disturbing the ability to interoperate at the *other* layers – as long the respondent did likewise. Once the OSI reference model was accepted, interoperability only required each layer be filled with standards.

Within the OSI model, the physical layer deals with the basic encoding of information within a physical medium (for example, how fiber-optic signals are to be read as bits or how Ethernet systems implement carrier sense and collision detection). Protocols of the data-link layer (such as the use of error-correcting codes) ensure that the bits are transmitted correctly. The network layer contains addressing information. The trans-ort layer governs connections (for example, how to disassemble, reassemble, and resequence transmissions in terms of component packets). The session layer dictates how two nodes can coordinate messages between them (for example, how to resynchronize an interaction that is terminated unexpectedly). The presentation layer allows the two to negotiate how messages are formatted (for example, whether ASCII is used for transmission). The applications layer contains basic operations such as e-mail, remote logins to other computers, and file and hypertext transfer.

It sounds complicated, and it was. The standards that fit the OSI reference model were spawned as expected albeit somewhat later than hoped, but they were not well supported by the major vendors and even less frequently requested by customers. The Internet, with similar but not entirely compatible standards, overtook OSI in the early 1990s. Today and for the indefinite future, it is the syntactic basis of cyberspace. The Internet's standard suite ignores the physical and data-link layer as being of interest only at the local area network (LAN) level. IP is the Internet's network layer. TCP covers its transport layer and a little bit of its session layer; because of such boundary crossing, Internet standards did not fit into the OSI rubric. The Internet lacks a capability to negotiate presentation syntax.[12] Finally, the Internet has a suite of application-layer standards such as SMTP for e-mail, and FTP and HTTP for file and hypertext transfer, respectively. See Figure 8 for an overall comparison.[13]

The top one-and-a-half layers of cyberspace, semantics of the real world and pragmatics, are rarely processed by machines. They are interpreted by people who read the content, find its meaning (the semantic layer), and infer its purpose (the pragmatic layer). Some applications, such as process-control systems, do process semantic information (for example, references to certain parameters of machine tools), but the semantics refer

---

[12] Nevertheless, the Internet supports its own encoding conventions (notably Multimedia Internet Mail Extension [MIME]) and refers to others, such as UNICODE for representing alphabetic characters, GIFs and JPEGs for picture encoding, and cryptographic standards.

[13] To clear up some confusion, "physical" layer standards in OSI refer to how signals are encoded for physical media such as wires and the radio-frequency spectrum. It does not refer to the existence of connectivity as such (in other words, it is not equivalent to the physical/phonology layer of the chapter's four-part schema).

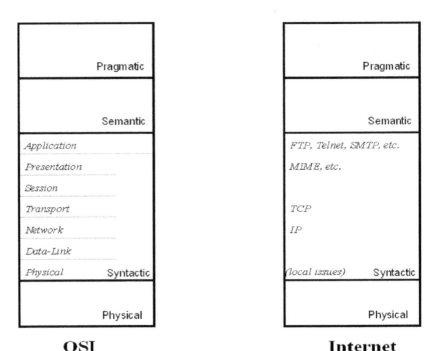

OSI   **Internet**

Figure 8. The Linguistics Analogy: OSI and the Internet Compared

to attributes engineered into what is being controlled. Systems would have to process semantic information based on their real-world content and knowledge rules – for which XML and software agents are harbingers – before the semantic and pragmatic layers of cyberspace come into play.

10.2 Complexity Facilitates Conquest

The complexity of today's information systems is a central factor in making them vulnerable to hostile and friendly cyberspace. Complex systems work at all only because they can differentiate among layers. The Internet, for example, is famously flexible precisely because the various packets can travel over radio frequency, twisted-pair, coax, fiber, or any other physical connections. But in practice, it is difficult to maintain boundaries, notably between the syntactic and semantic layers.

With systems, as in life, form has a way of dragging function along with it. Pace Marshall McLuhan's "the medium is the message," the way that information is presented is, in fact, its affective, and thus effective, content. Thus radio, in his view, was a hot medium that excited people while television was a cold medium that permitted people to distance themselves from what was going on. The same holds for food: how it is served ought not affect national diets – yet satisfying the desire to eat on the move favors the introduction of handheld food, and is thus an influence in favor of binding materials (such as crusts that hold together), which in turn have a high fat content. Many architects and city planners believe that the layout of a building or a city building influences the quantity and quality of social interactions within[14]; RAND's former 1950s-era headquarters had an open grid floor plan designed to maximize the number of random hallway interactions.

10.2.1 Complexity and Hostile Conquest

Maintaining a clean gap between a system's syntax and its semantics ought to be central to its design. Separation simplifies interfaces so as to help make large programs work.[15] But conquerors in cyberspace often succeed by finding where the barrier between syntax and semantics has worn through. One of the more common paths to computer mischief, for instance, is for a hacker to induce a buffer overflow: the computer expects an input value of a certain length; a much longer string of characters is entered; the initial string is processed semantically but the excess characters fall into memory, whereupon the computer tries to run them as process code (syntax). In a macro-level virus (such as the Kournikova worm), a sequence of commands (syntax) can be embedded in what is ostensibly an image file (semantics). A user's computer can be subverted by downloading rogue Web pages (notionally, just semantics) because

---

[14] Jane Jacobs, *The Death and Life of Great American Cities*, New York (Random House), 1961.

[15] Although some programs hard-wire data (such as conversion factors) into their algorithms, the separation of data and programs is a generally accepted practice. That noted, at least one computer language, LISP, ignores such distinctions completely.

the latter can contain instructions, such as Java code or, worse, Microsoft ActiveX commands. Time was when no one could receive a virus through e-mail, but modern client e-mail systems sometimes execute the instructions (such as macros) they contain by default. Until phone companies adopted out-of-band signaling starting in the 1980s, the same waveforms that carried conversation also carried switching instructions in the 2600 Hz band.

It should be possible to restrict interactions to known exchanges of semantic information, as traditional e-mail does. But this collides with trends that favor the ability to send instructions with data,[16] remote maintenance and updating, and the evolution toward ever-more detailed protocol exchanges (such as port specifications) for establishing network connections.

The difficulty of keeping the syntax–semantic barrier clean suggests why complexity spells the difference between a system that could in theory be secured, and one that in practice is not.

10.2.2 Complexity and Friendly Conquest

Crossing the syntax–semantics gap in the other direction can facilitate friendly conquest in cyberspace. Systems are complex – and so are the standards that must be adopted and, in turn, adapted to in order to facilitate interaction with a complex system. Whenever various systems must work with each another, the more influential ones are those that establish interface standards for the others. The subordinate system may, through co-evolution, work more easily with the master system, for reasons ranging from standards (syntax) to the way that they translate the world into categories (semantics). So, ostensibly syntactic causes (that is, how information is manipulated) produce semantic effects (that is, what information is collected, forwarded, manipulated, and so on). These effects can be overcome by the sufficiently determined. Yet, as system-held

---

[16] Examples include text marked up with HTML and then accompanied by Java and Javascript instructions or objects that seamlessly mate data structures and procedures to manipulate them.

knowledge and logic increasingly affects the information content that decisions are based upon, and as systems grow more complex and therefore more hostage to their own syntax and semantics, these obstacles will grow larger and thus more decisive. Unless users are willing to accept less choice or fewer tasks they can execute, complexity will remain a fact of systems life.[17]

Complexity lends standards – and thereby those who controls them – much of their influence. In the 1980s, Microsoft's DOS dominated computers, but that fact alone did little to boost Microsoft's applications in the market. DOS was comparatively simple and applications called on it in a limited number of ways. Microsoft's Windows 3.1 was much more complex and intrusive. It determined the look and feel of all subsequent applications, which had to call on the operating system far more often and in many different ways. As Windows overtook DOS, Microsoft achieved majority market share in applications – word processing, spreadsheets,

---

[17] The late Michael Dertouzos had been eloquent on the difficulty of using computers. Nevertheless he was among the optimists in believing that the problems are not inherent in computing as such but result from design choices that reflect the tendency of engineers to talk primarily to other engineers. See, especially, Michael Dertouzos, *What Will Be: How the New World of Information Will Change Our Lives*, San Francisco (HarperEdge), 1997, chapters 12 and 15, in which he argues:

Unfortunately computers and communications networks are not easy to use. The manual for a word processing program is as thick as a dictionary. Even telephones have become complicated not to mention inhuman, like automated corporate answering systems that force us to suffer through tedious push-button choices before letting us talk to a real person – if at all. The most important property of an infrastructure – the ability to make possible numerous independent activities – is not met by today's information infrastructures either. Surely, individual computers do support many useful applications, from spreadsheets to CAD. But they cannot perform easily thousands of different tasks over a network. My computer cannot find me the car with the greatest headroom because different manufacturers keep their data in different forms and on different sites. This is the norm today. Different machines and different software packages use different rules. You must stand on your head and use all kinds of arcane mechanisms to make any sense out of them. Browsers and the Web don't help in this regard because you end up doing an inordinate amount of work searching without any assurance as to the outcome (p. 16).

See also idem, *The Unfinished Revolution: Human-Centered Computers and What They Can Do for Us*, San Francisco (HarperInformation), 2000. Since then, at least one site, edmunds.com, has gone to the trouble of collecting such information about cars and putting it in standard format for its customers' use.

graphics packages, and databases – where it had hitherto been second or worse. Mastering Microsoft's more complex interface – which no one did as well as Microsoft – suddenly became a lot more important to having an efficient and stable program. Microsoft has been accused (and convicted) of leveraging its command of operating systems to dominate downstream software such as browsers and other applications. For a while, there were intense debates over whether its ".Net" initiative (as section 10.4 discusses) was a ploy to leverage its software monopoly into a pole position in cyberspace, but ".Net" seems to have morphed into its next generation of operating systems and programming tools rather than representing a leap into cyberspace. The more intriguing and chancy leap is from syntax to semantics directly. Can Microsoft leverage its software skills (a classic syntax endeavor) to become a content provider on the Web? It is unclear whether what modest success it achieved has anything to do with its leverage or more to do with its very deep pockets and generally talented workforce.

One could dismiss standards as a design choice among comparable technologies: should plugs be flat or round? But standards can also define the architecture of cyberspace by making certain actions easier and others hard. Because the Internet relies on packet and not circuit switching, more intelligence can be put in the periphery rather than in the switching fabric,[18] the system as a whole is more decentralized, and it is easier to support the high and variable bit-rate requirements of multimedia. Conversely, event-based administration, usage-based billing, certain security features (such as resistance to flooding attacks), and priority channels are harder to implement. The programming languages Ada and C are analogous ways to convert algorithms into assembler language. Yet, Ada, with its strict rules and type enforcement, empowers project managers, whereas C, with its concision and freedom, empowers programmers.

If systems had perfect security, hostile conquest in cyberspace would be impossible. If systems had perfect transparency and interoperability, friendly conquest would lose at least one vector of attack – but neither should be expected anytime soon.

---

[18] Isenberg, op. cit.

10.3 Semantics

Semantics can be divided into the real and the abstract. Real concepts refer to the tangible world. Abstract concepts include mathematics (such as CAD structures), grammatical structures (such as paragraphs), well-accepted symbologies (such as for music or chemical structure notation), and computer-based processes (printer and machine tool commands, for example). Many of the latter take meaning only within the context of a specific system (such as machine tool instructions or the logical characteristics of a network printer).

Of the two, abstract semantic interoperability is farther along, largely because what is to be standardized has already been abstracted, and often in common ways. In other words, a printer driver relies on the fact that the printer's functions have already been turned into ones and zeros; what is left is to establish a correspondence between what the computer wants the printer to do and how the printer manages to do it.

Real semantic interoperability, such as what "street" means, lags behind. Interoperability conventions at the physical, syntactic, and abstract semantic level (for example, conventions for designating italics) combine to send someone a Web page intact. Yet, its content is expressed in human language for humans to manipulate. Machines display and humans read. The same holds for pragmatics (for example, "Why was I sent this e-mail? What was the sender's motive?"). It is inferred from and not transferred within a Web exchange. Operations and thus interoperability in cyberspace currently extend two-and-a-fraction levels and then people take over.

In 1998, a standard syntax, XML, was invented[19] by the Worldwide Web Consortium, led by Tim Berners-Lee, the Web's inventor. Its purpose is to denote and convey tags; such tags could carry semantic content. XML is thus a syntactic convention that permits semantic interoperability. To pass information for machines to process requires tag set definitions that are uniform, at least across specific domains (astronomy, for example). Blizzards of tag sets standards have started blowing in, a large share from the world of business, where accounting standards and contract laws

---

[19] More precisely, XML was refined from a 1986 standard, SGML.

already serve to categorize activity in standard ways – and there are still disagreements over them. The successes and setbacks that followed prove anew the old saw about standards: how nice that there are so many to choose from.

The advent of semantic standards would be a sharp shift in computing. Fuzzy words in human language would assume a precision, perhaps a false precision, in computer language. Once standardized, even if only within domains, categories that hitherto supported generalization for individual logic processing programs (for example, "If this is a street, then look both ways before crossing it") could graduate to become a basis of generalization for every program shared across such domains. Some semantic standards describing real-world concepts antedated XML. Starting in the late 1980s, for example, terms from the ANSI X12 standard for electronic data interchange (EDI), such as price, quantity, and shipping date, were understood by people and automated scheduling systems in common ways even though they covered heterogeneous things. The geospatial community has standard terms for commonly encountered macroscale entities (such as mountains). DoD's data dictionary attempts to do this across the military. Such terms have all been reformatted for XML conformance.

XML may permit complex information to be passed as semantic content in ways that frustrate at least one path to hostile conquest. To explain why, consider object-oriented programming. Objects are collections of data structures and the delimited set of procedures that work on them (in other words, no procedure that was not predefined could be applied to data so structured). This eliminates certain kinds of errors[20] and permits higher levels of abstraction. Two parties, for instance, may want to build a simulation over the Web using synthetic creatures with particular behaviors. A "dog" may be one object, a "horse" another. Because they are both animals, they inherit the common characteristics of animals such as age, health, and number of legs. But they also differ: dogs play frisbee and horses are ridden, but not vice-versa. Someone who would contribute a

---

[20] Using object-oriented programming, for instance, prevents one from adding the years 1991 and 1992 (resulting in 3983) even though they are both numbers.

horse of a particular description (to populate a simulation, for example) would have to package the description with the procedures that apply to horses. Such procedures, in turn, may or may not be trustworthy, either because the sender is malevolent or, more likely, because not all bugs have been squashed. XML, however, permits one to send only a description of the standard parameters of a <horse> ("<" and ">" are how XML denotes tags). The recipient, in turn, can refer to a standard, an authenticated and publicly available set of procedures associated with <horse> and stored in a specific Web repository. Indeed, there could be many different procedures associated with <horse> defined in many different places; fortunately, there are metastandards to indicate the standard to which the procedure is referring.[21] The procedures in depositories, because they are meant to be referred to and used by many people, would likely be scrubbed and certified as safe to use.

Using the word "horse," especially if defined as an animal, would invoke the explicit assumptions similar to the implicit assumptions of human language. Consider the following sentence: "A <horse> was <spied> in <downtown Cleveland> on <New Year's Day 1999>." This could be added to some public stock of knowledge about horses and ultimately used to support inferences. References to horses would, to a computer, become references to such inferences. Some of these inferences may in turn be private to the recipients, just as the word "horse" evokes different connotations among different people. Yet, what two people call horses – the denotations – should be the same; for example, whether or not ponies are included should be decided upon consistently.

So, semantic standards could reinvent language – for computers – but they would have to make the distinctions that all definitions do. More than with human language, the distinctions made by the availability of terms would dictate which generalizations can or cannot be supported. To use a gross example, lumping "terrorist" and "guerilla fighter" together and then generalizing that category would lead to assignments (designating a

---

[21] XML namespaces and the broader RDF are described in the "Resource Description Framework (RDF) Model and Syntax Specification," www.w3.org/TR/REC-rdf-syntax/.

group as "terrorist," for example) and rules that would differ from those that would result if "guerilla fighter" and "soldier" were lumped together as "warfighters." How much difference, in what direction, and to what extent differentiations are overcome by making finer distinctions to fit evidence are open questions. These definitions may be supportable but rarely objective and not necessarily neutral. And computers are likely to remain well behind humans in questioning these categories in the light of their "experience" – that is, collected facts that would feed their knowledge base.

Might cyberspace support a soft form of the Sapir-Whorf hypothesis in the sense that semantic standards can affect how computers and their programmers think? The answer may depend on the sophistication and complexity of the processes that use standard terms. There is no a priori reason that artificial intelligence cannot ultimately be as supple as natural intelligence; if so, it can surmount whatever barriers to cognition are imposed by how things are categorized (that is, tag sets). Complexity, however, may lead to inflexibility in that the more that certain rules rely on a definition, the harder it is to change them. They are either used as is or changed wholesale, a task much more all-or-nothing than is true for everyday human cognition. Although the Sapir-Whorf hypothesis, like many rules about cultural development, is hard to test, the emergence of semantics for information processing has the potential to build language afresh.

To the extent that the Sapir-Whorf hypothesis applies in cyberspace, those who can craft the standard semantics may have a tool to channel cognition in specific directions. The crafters need not be all-powerful; the tendency of standards to gain universal acceptance because they are becoming dominant and everyone jumps on the bandwagon is good news for those with luck, timing, and position.

Conversely, Berners-Lee has argued that common semantics may be inferred, in part, through analytic engines that can comb the Web and see how terms are used in practice.[22] In what he calls the Semantic Web, the group process will dominate competing de jure or de facto

---

[22] See Tim Berners-Lee, *Weaving the Web*, San Francisco (Harper San Francisco), 1999, pp. 177–96.

standards.[23] There will be few avenues for unwarranted influence through the imposition of standard semantics.

10.4 Pragmatics

With semantic understanding off in the future, speculations on a pragmatic layer may seem quite premature – but such speculations suggest yet another reason why the power of hostile conquest may fade before the power of friendly conquest.

Systems might be able to use pragmatics if they are built from agent-like components from different owners or developers – much as human societies are built from independent entities under no master control. Integration is more than interoperability, just as effective organizations require more than that their members speak the same language. Heterogeneous databases may require interoperability before they can update themselves in a synchronized fashion, for instance, but the choice of *which* databases to update based on changes in what other databases has to be determined first. Before refrigerators talk to grocers about their milk supplies or cars talk to mechanics about funny noises,[24] both the grocer's and the mechanic's computer systems need to anticipate the possibility of a call and know how to respond to such calls in ways expected by the refrigerator and the car, respectively.

In a closed system, the participation of a component is built into its programming, largely as a result of dividing complex problems into specific tasks. But this presumes that each component answers to the same master and that all of the conditions under which each operates can

---

[23] See Tim Berners-Lee, James Hendler, and Ora Lassila, "The Semantic Web," *Scientific American,* May 2001, pp. 34–43. As *Business Week* reported in 2002:

By 2005 [Tim Berners-Lee] hopes to begin replacing the Web with the Semantic Web – a smart network that will finally understand human languages and make computers virtually as easy to work with as other humans (Otis Port, "The Semantic Web," *Business Week,* March 4, 2002, pp. 97–102).

See also Andrew Updegrove, "The Semantic Web: An Interview with Tim Berners-Lee," www.consortiuminfo.org/bulletins/semanticweb.php, June 2005.

[24] An example cited by David Roddy, vice president of VerticalNet, quoted in Robert Hof, "Look Ma, No Hands," *Business Week E-Biz,* November 20, 2000, p. 132.

be prespecified. Even in such cases, distributed negotiation may work better.

Take, for instance, the problem of coordinating sensors to provide good spatial, spectral, and temporal coverage[25] of a battlefield – a problem that may have to be wrestled someday in reifying the GIG.[26] If bandwidth is limited, local coordination among sensors will be needed to govern spectrum contention. If only a few designated sensors were provided with the energy resources to transmit information, the sensors must also coordinate among themselves to fuse and otherwise preprocess sensor readings first before passing them on for retransmission over thin links to a receiver outside the battlefield. Ground sensors will have to be coordinated to work with sensors on UAVs. Such coordination cannot necessarily assume that the position and parameters of each sensor are known in advance. Inevitably, some sensors will blow off course, get stuck, point the wrong way, or be disturbed by animals, passersby, and even the enemy.[27] To helps its cohorts figure out where it is, each sensor, once deployed, might announce what it is assigned to be (for example, collectors, relays, or fusion nodes its location), what it is looking for, where it is looking it, and which spectral bands it is using. Each might negotiate with each another for bandwidth. One pattern of organizing sensors may be optimal for detecting certain phenomena (such as an expected tank movement) even as another is optimized for different tasks (such as a localizing troop movement). As the battlespace changes, each sensor might ping others to

[25] DARPA, for instance, has funded Berkeley to develop "smart dust" to meet such needs (http://robotics.eecs.berkeley.edu/~pister/SmartDust/ is a portal into the program). See also Brendan Koerner, "Intel's Tiny Hope for the Future," and Martha Baer, "The Ultimate On-the-Fly Network," both in *Wired,* December 2003, pp. 236 and ff.

[26] The Global Information Grid – called a "system of systems" – is more like a "federation of systems," in part because it contains pieces from every armed service, many defense agencies, and, sometimes, allies. See Annette Krygiel, *Behind the Wizard's Curtain,* U.S. DoD C4ISR, Washington (Cooperative Research Program), 1999, especially Chapter 2, "Systems of Systems and Federations of Systems," pp. 31–47.

[27] See, for instance, G. L. Pottie and W. J. Kaiser, "Wireless Integrated Network Sensors," *CACM,* 43, 5, May 2000, pp. 51–8; Harold Abelson et al., "Amorphous Computing," *CACM,* 43, 5, May 2000, pp. 74–82; Glover T. Ferguson, "Have Your Objects Call My Objects," *Harvard Business Review,* June 2002, pp. 138–44; Jean Kumagai and Steven Cherry, "Sensors and Sensibility," and subsequent articles, *IEEE Spectrum,* July 2004, pp. 22–48; Michael Kanellos, "Making Sense of Sensors," http://news.com.com/2008-1082_3-5829415.html?tag=nefd.ac, August 15, 2005.

determine whether reassignment is called for, to elicit the capabilities of its neighbors, to correlate what neighbors see against what it sees, and to inquire further if anomalies are found. When cued, some sensors might switch parameters (for example, field of view versus acuity, or cue versus pinpoint) or turn certain receptors on and others off. Sensors may be tasked to zoom in on a target. Sensors may need to be told when to be silent or when to switch frequencies. Rich dialogue will be needed for all this.

Pragmatics enters the equation whenever various components are fielded without a priori basis for knowing how their counterparts are tasked or how they are doing. A comparison can be made with software agents, an agent being a program that interacts with the rest of cyber-space to satisfy its owner's need. Shopping "bots," for instance, are agents that pull up Web pages to discern an item's key characteristics and its price. More sophisticated agents start with ill–defined requests and search through knowledge spaces. Human dialogue has ways to deal with differences between information that is asked for and the information that is really needed, as this dialogue suggests:

– Is it raining?
– *Why do you ask?*
– So I know whether to wear a jacket. (Rather than, for instance, to determine whether plants need watering.)
– *We are not going anywhere.*
– Don't we need groceries?
– *Oh, I picked them up yesterday.*

Here, divining why a statement was made or a question asked is necessary to know how to respond. One side knows what information it has; the other side knows what it needs to have, but without negotiation each can only guess what to pass to the other.

Or, take another vignette. Two software agents, Alice and Bob, are negotiating over buying a machine tool. Alice asserts it needs a tool to perform certain functions while meeting ancillary requirements (such as safety, training, and maintenance). Bob replies it has an item with capabilities as measured by certain parameters. Assume that they share a common semantics (for example, for Alice to specify the need to drill through steel of a certain American Society for Testing Materials [ASTM] rating, Bob must understand "drill," "steel," and what tests are implied by such a rating). To conduct the dialogue, they need terms to define transactions – not only query and response, but also assertions, assumptions, and alternatives. Alice wants to know how much the drill's operation shakes its substrate and thus nearby machinery (all of its machinery is bolted to a common metal frame). But to be meaningful, its context must be made clear; for example, vibration (and in what dimensions, and under what circumstances, and so on) may depend on how it is used. Has Bob, most of whose other customers bolt machinery to cement floors, measured this feature? What sort of reply would assuage Alice's concern? Alice may ask what parameters remain to be asked about.[28] Once a close enough match on the object occurs, each side needs to agree on whether negotiations proceed to other facets (such as delivery, price, and guarantees) sequentially or simultaneously. The more that each side employs a common negotiation model, the less the need there is for an explicit transfer of requests. But at some point, they must convey metanegotiation information as well; each side's response will be off if Alice expects a fixed-price model and Bob uses a more bazaarlike process.

Both examples involve negotiations – the conveyance of purpose, goals, and context. They contain not only responses to queries but also background. Intelligent negotiations are those in which possessors of information can use what they know to satisfy the needs that others have. In the preceding example, the machine tool seller should know not only what someone is asking for but also why: what is the purpose of requesting certain information? Thus, pragmatic lexical primitives (although pragmatics need not be expressed in words as such) would cover queries, responses, and metaresponses (for example, whether an answer is incomplete and why). Other concepts include assertions of need to know, urgency for the information, format negotiations, state

---

[28] Unlike humans, Bob could simply transfer everything it knows to Alice and let Alice query such information itself without the constant back and forth. So, why not do this? First, many of Bob's answers may actually be information that cannot be retrieved without polling others. Second, Bob's owner may not want to transfer any information except in the context of a negotiation that bears some promise of profit and within the context of a particular trust model. Dialogue minimizes the information that has to be released.

declarations (for example, whether a query is fresh or a follow-up), and resource constraints. Respondents should be able to express and describe the certainty, quality, authenticity, and, perhaps, rationale for their estimates, contexts in which the answer makes sense, requirements for further clarification, alternative or complementary data sources, assertions (for example, "I intend to do this"), rejoinders ("do that instead"), taskings, tasking authority, tasking justification ("I sense this, so do that"), and metrics for task completion (so that the taskee knows whether to continue or quit). Pragmatics provides a common lexicon for "why is this so?," "why does this matter?," and "what problem will the answer solve?"

The ability of an information system to exhibit skepticism about source, content, and even motive – communications intrinsic to a pragmatic layer – would permit systems to filter out more bad messages. Computers that recognize and carry out the *intentions* of their owners, if they can do so reliably, may be able to defeat attempts to get them to act at odds with their basic directives.[29] Dialogue may provide yet more techniques for building secure integrated systems from insecure components. It does require machines to exhibit skepticism and self-consciousness – at least in the sense that it maintains some state of knowledge that can link its goals to the effects of its actions. In that sense, skepticism is a filter much as passwords are.

How far has the market gone in supporting pragmatics with tools and standards? In 1999, Hewlett-Packard tried to market its concept of E-services. In one advertisement, a car on the verge of breakdown automatically notifies the nearest mechanic, which e-mails back to the car a diagnostic of the problem and writes a purchase order for the required part. The driver would be blissfully unaware of the sequence until directions for reservations at the nearest hotel would show up on the automobile's GPS-based map. This capability requires systems to recognize each

---

[29] Readers of Isaac Asimov's *I Robot*, Garden City (Doubleday), 1950, will recall the Three Laws of Robotics:

1. A robot may not injure a human being, or, through inaction, allow a human being to come to harm.
2. A robot must obey the orders given it by human beings except where such orders would conflict with the First Law.
3. A robot must protect its own existence as long as such protection does not conflict with the First or Second Law.

other as entities in order to conduct negotiations to meet their ends. HP, unfortunately, had lacked sufficient depth in programming, networking, customer relationships, and, perhaps most importantly, prime real estate in cyberspace to move its idea much beyond vaporware.[30]

Microsoft had no such constraints. It announced its ".Net" initiative shortly thereafter, and thereby launched its foray Web Services. Dialogue would expand from want-data-have-data to the ability to define products and services, find suppliers for them, establish a way to discuss terms of trade for them, and make and satisfy attendant requests of each other. This is no small ambition; it requires everyone, or at least everyone within a given domain, to share a common pragmatics vocabulary. With proposed standards have come the usual standards battles – petty in behavior and enormous in impact – this time pitting IBM (and its WebSphere program) and Microsoft against Sun Microsystems, Java's champion.[31] Companies that would negotiate with each other first have to find each other using, as currently contemplated, Universal Description, Discovery, and Integration (UDDI) and Web Services Description Language (WSDL). They would send each other messages using Simple Object Access Protocol (SOAP) as the transport mechanism and conduct these negotiations using terms specified in yet another standard, ebXML (electronic business XML). These negotiations could be programmed using Microsoft's Visual Studio ".Net" and processed using Microsoft's ".Net" server – or so the theory goes. Microsoft's programming tools, and its historic and current expertise, have sold well; ".Net" servers have come more slowly, and the premise that negotiations would be dependent on Microsoft-dominated cyberspace has met market resistance.[32]

---

[30] Charles Cooper, ".Net and the Emperor's New Clothes," http://news.com.com/2010-1071-948117.html August 2, 2002.

[31] In this version of the soap (vice SOAP) opera, IBM, Microsoft, et al. have formed a Web Services Interoperability (WS-I) Organization to create Web services standards. Sun, Java's inventor, wanted in as a founding member on the strength of its historic contributions, but was informed that, in effect, it could have a seat in the General Assembly but not as a permanent member of the Security Council.

[32] Nevertheless, Miguel de Icaza's Mono project created an open-source set of hooks to interface with ". Net". See ".Net inches closer to fruition," *Infoworld*, www.infoworld.com/articles/hn/xml/02/07/22/020722hnoreilly.html, July 22, 2002; David H. Freedman, "Sellout or Savior," *Technology Review*, September 2004, pp. 44–9.

10.5 Lessons?

Over time, as information technology has advanced, it has replicated the layers that have long characterized human speech. This is clearly so for the first two-and-a-half layers: the physical, the syntactic, and the lower half of the semantic layer. The advent of XML and the activity entailed in defining tag sets suggests that the upper half of the semantic layer may be filled in over the coming decades.

Unlike human language, computer language requires that the various layers be considered as complete and independent. Indeed, the OSI reference model requires the seven sublayers of the syntactic layer be independent of each other. But systems are complex and growing more so. Unintended violations of independence permit certain types of mischief. Complexity associated with standards, in turn, give scope to dependence and hence friendly conquest in cyberspace.

The advent of a true semantic layer is likely to create more avenues for friendly conquest. The influence accrues to whoever wins the right to define the words that refer to the real world – that is, what they encompass and exclude. Words express categories; categories, in turn, are the basis for rules and tenets. Once these rules are embedded in enough programming, they reinforce the semantic categories, perhaps more powerfully than they do in human speech. People, after all, are more flexible than machines.

A developed pragmatic layer, should it occur, would offer even more impetus for friendly conquest by reinforcing the role of semantic exchange. Conversely, pragmatic primitives of information exchange could provide, in machine form, the means by which systems may express skepticism about communications and request clarification, bolstering their ability to resist manipulation in ways that people already do now.

11

Managing Conquest in Cyberspace

Cyberspace lends itself to policy questions of the sort applicable to other media. How can one best use power in this medium to do unto others? How can one best maneuver in this medium to prevent being done unto? Insofar as conquest in cyberspace has hostile and friendly vectors, these two policy questions turn out to be four:

- How can the method of hostile conquest be used wisely?
- How can countries ensure that others cannot conquer those parts of cyberspace they deem worth protecting?
- How can a country best pursue its interests by exercising friendly influence through cyberspace?
- How can a country best resist the friendly accretion of unwarranted influence in cyberspace by others?

The first question can be interpreted as one of command and control, strategies, operations, techniques, and tools. Information warfare is not without its conceptual problems, such as the ability to define legitimate targets – but so are other domains.

The second question requires first determining what it is that must be defended and why. No one doubts that it is a government responsibility to defend Pittsburgh against Iranian missiles or chemical clouds; the good burghers are not expected to take this upon themselves. But, in cyberspace there *are* limits on what governments should be expected to contribute the common defense.

The third question begs: what ends would be served by seduction and how could one know such ends were met? If the point of seduction is to bind others to us, at what point does the reverse hold – that we have become bound to them in return?

The fourth question calls for understanding what it means for influence to be unwarranted, why it is undesirable, and what responsibility the U.S. government has to mitigate its effects. The U.S. legal code does discourage many forms of influence, especially when wielded by foreigners (for example, foreign lobbyists must register themselves as such). There are limits to who can own how many of which media outlets in any one city. But, as often has been noted, the Internet may make irrelevant or unenforceable many regulations based either on scarcity (of spectrum, for example) or on location.[1]

11.1 Conducting Hostile Conquest in Cyberspace

As noted, especially in Chapter 4, the most important aspect of the command and control of hostile conquest operations is the difficulty of exercising it. Unlike, say, tank warfare, where private individuals tend to lack the requisite equipment and rogue tank units are quite obvious to all, the tools of information warfare are as ubiquitous as fingers and keyboards. Individuals and small teams may be quite effective if well informed, clever, and lucky. Success at hacking requires daring, initiative, and the rapid identification and seizure of fleeting opportunities. The best hackers are independent and all too high-spirited; it is an activity unsuited for plodders. This hardly constitutes the easiest crew over whom to exercise control, much less chain of command. Worse, intelligence preparation of the battlefield – a must – requires hacking prior to having the information that would justify permission to do so. Unfortunately, the distinction between breaking and entering to scope the joint and doing so to trash it is awfully thin.

---

[1] Whereas one's species may be impossible to determine in cyberspace, location is not so difficult anymore. Well over 90 percent of the time, a clever server such as those employed by search engines can determine where the client sits. Without such capabilities, localized advertisements would be almost impossible. To test this proposition, type a vendor category (such as "dentist") into a search engine and look at whose advertisements accompany the response. Countries such as China, with its limited external gateways, have no problem figuring out who is inside and who is outside the Great Firewall.

Equally thin is the distinction between legitimate targets of war and off-limits civilian targets. The same holds for distinctions between conventional forces (fair game) and nuclear forces (perhaps a router too far). Any success by information warriors against nuclear establishments may persuade victims that they could lose control over weapons before they get a chance to use them – with unfortunate consequences for all concerned. Furthermore, others may not necessarily respect the same firebreaks observed by the United States between nonviolent conquest in cyberspace and violent conquest in other media. Thus, compared with conventional warfare, command and control over information warfare must be exercised with far more attention. The specter of some president who lets others program his VCR but picks cyberspace targets himself may seem contradictory, but if anything goes seriously wrong, the buck will rise up the chain at a head-snapping speed.

Evidence from the war over Kosovo suggests that the Pentagon was skittish about giving hackers in the field free rein. Proposals to go after banks that held the money of Sloboban Milosevic's friends or go after civilian infrastructures were put to DoD's lawyers, who concluded that the same rules of war that governed destruction in real space could be applied to targets in cyberspace.[2] Thus, many of the contemplated actions, as long as their effects could be controlled, were technically acceptable. Clearly there was no useful moral or legal distinction to be made between hacking into a military computer and using other means to force its reversion to silicon dust. But, also in the end, very few information operations were tried in that conflict, or at least publicly admitted.

The practical matter of determining the odds of causing collateral damage should give pause, if nothing else does. Collateral damage following an information attack is even difficult to ascertain after the fact. If U.S. corporations do not go running to the FBI every time they find some hacker taking excessive liberties with their enterprise infrastructure, then

---

[2] Department of Defense, *An Assessment of International Legal Issues in Information Operations*, Washington (Office of General Counsel), 1999. The report argued, "there is no requirement that an act of self-defense use the same means as the provocation, that the object of the attack be either a similar type of target or the means used in the offending attacks, or that the action be taken contemporaneous with the provocation, particularly if the attacker is responding to a continuing course of conduct" (p. 19).

why assume that foreign victims will? Damage in cyberspace may also spill over into the domains of neutral countries. Many of the bank accounts held by friends of Milosevic were in Greek or Swiss banks. Should one of DoD's hackers have succeeded in zeroing out their bank accounts, the account holder's exchequer may be unexpectedly depleted – and the bank's exchequer unaccountably incremented. But, in the end, there may be more lasting damage to the bank's reputation than harm to the victim. A bank deposit, after all, is a promise on the bank's part to repay money. This promise is not voided by the inability of the lender to remember the extent of the debt, but errant sums may cause some to question the convenience of the bank's memory. Worse, the more the world's economy becomes globalized, the harder it is to predict the ripples from any one act of mischief.

Although the cautions on collateral damage are always in order, at least the *intended* effects of hostile actions in cyberspace against terrorists are more broadly acceptable. Authorities have pressured service providers to shut down access. British intelligence services, for instance, have reportedly attacked the Jihadist presence on the Internet to make it more difficult for sympathizers and potential militants to access Jihadist propaganda, training videos, and ideological writings.[3] Whether it is because such efforts have succeeded or merely have been anticipated, the Jihadist presence on the Internet has already assumed evanescent qualities. Some instructional or propaganda materials have been placed on innocent sites (such as the State of Arkansas Highway and Transportation Department) that hackers have managed to finagle access to (a trick well known to free-lance pornographers). Potential recipients are then e-mailed the site's Web address. Often they must move quickly before the address is compromised (less than a day in the case of the Arkansas site) and other cubbyholes must be found. One site, Alneda.com, migrated from Malaysia to Texas, and then to Michigan over the course of a few weeks.

To date, it is unclear how much good has been accomplished by keeping terrorists on the run in cyberspace. Given how little officials understand

---

[3] Shawn Birmley, "Tentacles of Jihad: Targeting Transnational Support Networks," *Parameters*, Summer 2006, p. 40.

about Jihadist terrorism in general and the insurgency in Iraq in particular, a careful monitoring of Web sites can only help increase people's
understanding of what drives the Jihadists.[4]

The inability to find permanent sanctuary on the Internet has not pre-
vented terrorists from finding virtual sanctuary there, and some Jihadists
have become quite sophisticated about covering their tracks. As long as
there are some (of the world's millions of) sites they can hijack, and as long
as mostly everyone they rely on has someone else's contact information
in cyberspace, a network exists. At the speed of light, a little inefficiency
may not be noticeable. Except for communicating to those off the Jihadist
network altogether, such terrorists do not seem to need much permanent
turf.

If disruption has its limits, this leaves two other avenues for offensive
operations in cyberspace against terrorists: exploitation and corruption.
Of exploitation, there has certainly been a great deal, both by intelligence
agencies and by free-lancers[5] and at least one case in which such monitoring led to the apprehension of a suicide terrorist while he was still
alive.[6]

What the Internet seems also to have done, though, is to make it easier
for the more amateurish plots to be detected and arrested before they have
gotten very far – leaving many to wonder whether they had the potential
to get very far in the first place. Two examples from mid-2006 illustrate
this.

In June 2006, Canadian officials arrested seventeen and charged them
with conspiracy to carry out terrorism, with the Parliament in Ottawa
being the most likely target. It is believed that the Canadian Security
Intelligence Service was tipped off by British authorities monitoring the
Internet movements of Younis Tsouli, who, in turn, was contacting radical
recruits in Toronto and Atlanta in chat rooms.[7] The *Toronto Star* reported

[4] For instance, see the International Crisis Group, *In Their Own Words: Reading the Iraqi Insurgency*, Middle East Report No. 50, http://www.crisisgroup.org/home/index.cfm?l=1&id=3953, February 15, 2006.
[5] Even from Montana. See Blaine Harden, "In Montana, Casting a Web for Terrorists," *Washington Post*, June 4, 2006, p. A 3, available at www.washingtonpost.com/wp-dyn/content/article/2006/06/03/AR2006060300530_pf.html.
[6] Wallace-Wells, op. cit., p. 30.
[7] "Home Grown Terror: It's Not Over," op. cit., p. 20.

that in 2004, the Canadian intelligence agency began monitoring Internet exchanges, some of which were encrypted.[8] So cued, the government opened up a criminal investigation in early 2005. For months, police monitored the movements, and intercepted the e-mails and telephone calls of Fahim Ahmad, one of the seventeen. Internet usage was also how investigators found out the identity of two American citizens, now in custody in the United States, whose anti-American beliefs led them to cross paths with some of those in custody in Ontario.[9]

In 2006, Lebanese officials arrested an individual for allegedly conspiring to blow up what were probably transit tunnels under the Hudson River. Supposedly, FBI agents who had monitored Internet chat rooms used by extremists learned of the plot and inferred what the possible targets were after investigators pieced together code words from their conversations.[10] The Lebanese Internal Security Directorate said Assem Hammoud, going by the nom de guerre Ameer Andalusi, was initially noticed on an Islamist Web site used to recruit jihadis; he had sent maps and plans for an operation to other members of his group over the Internet. The Lebanese authorities located him based on the IP address embedded in his posts, which showed him to be in Beirut.[11]

How long investigators will be able to exploit such communications methods employed by terrorists is unknown. In response, such groups constantly search out harder-to-trace methods of communication, such as counterfeit Subscriber Information Module (SIM) cards or satellite phones that are dumped off after a single use.[12] Given the deserved paranoia in such circles, even the best interception techniques will be lucky to yield names, addresses, times, dates, and places.

---

[8] Ian Austen and David Johnston, "17 Held in Plot to Bomb Sites across Ontario," *New York Times*, June 4, 2006, p. A1.

[9] Anthony DePalma, "Ontario Terrorism Suspects Face List of Charges in Plot," *New York Times*, June 6, 2006, p. A11.

[10] Fox News, "FBI Busts 'Real Deal' Terror Plot Aimed at NYC-NJ Underground Transit Link," http://www.foxnews.com/story/0,2933,202518,00.html, July 7, 2006; Eric Lipton, "Recent Arrests in Terror Plots Yield Debate on Pre-Emptive Action by Government," *New York Times*, July 9, 2006, p. A13.

[11] According to Lebanese authorities cited in A. L Baker and William K. Rashbaum, "3 Hold Overseas in Plan to Bomb New York Target," *New York Times*, July 8, 2006, p. A1.

[12] "Home Grown Terror: It's Not Over," op. cit., p. 24.

The usefulness of corrupting the Jihadist presence is cyberspace is limited by the nature of terrorism itself. It is quite a stretch to characterize terror networks as finely tuned machinery that relies on the reliable and trustworthy exchange of unmonitored information (much as, say, a typical bank or oil refinery does). Terrorist operations, like Soviet military gear, are rough-and-ready, plotted and conducted with the intense awareness of the possibility of failure and accordingly robust against bit errors, if not always successful.

All in all, what makes the Jihadist terrorists so difficult to confront in cyberspace is that they have not attempted to conquer it in any way, shape, or form. They defend nowhere, the better to appear anywhere.

11.2 Warding Off Hostile Conquest in Cyberspace

As Chapter 2 argued, it is by no means certain that the security of cyberspace writ large will be anything but a notional threat. Perhaps computer security may be the norm, and consequential attacks rare and reversible. Were it so, policy would have little to do. Apart from a small admixture of law enforcement for after-the-fact prosecution, the defense of cyberspace would be a concern solely of those with systems to protect, much as health is largely a personal matter.

And that would be a good thing – because if the balance between careful software construction and business ambition or, more broadly, between good and evil within computerdom were such that systems were not reliable and information insecure or tampered with, finding an effective public policy would not be so easy. This can be illustrated by a series of five vignettes.

11.2.1 Byte Bullies

The five-part story to be told starts by assuming a growing din in cyberspace resulting not from crime but from war.

Strategic information warfare is rising.[13] This raises the cost of running all information systems and puts a pall on technology-driven growth. New devices are

---

[13] The term refers to R. C. Molander, *Strategic Information Warfare: A New Face of War,* Santa Monica (RAND), 1996.

viewed warily – how many more problems will they cause? The same holds for services (such as tele-medicine and home grocery shopping) that require exchanging large amounts of personal data for the purposes of customization; ditto for data mining, or the use of simulation to exchange product and process descriptions. Yet, economics is hardly half of the story. Shadowy outfits demonstrated that they had their hands on the volume controls of cyberspace and could therefore demand and receive favors for "protection." Suspicion rested upon one particular country. Not only were attacks persistent and sophisticated but institutions that demonstrated favor to the country seemed to avoid such problems ("seemed" because many of the favors and most of the faults are not made public). The United States government offered to help victims overcome the effects of hacking, but few took it up for fear of giving U.S. officials privileged and revealing insights into their compromised information systems. No one was entirely sure that their own systems were free of gremlins that could strike between the time when government assistance was accepted and when it could be effective. Most system owners, especially overseas, cut their own deals with that particular country, and enjoyed a measure of relative cyber peace as a reward.

The practical difficulty faced by defenders in this scenario is determining the source[14] of the mischief so as to convert the difficult problem of defending cyberspace into the well-understood problem of wielding conventional state power against a defined foe.

How might a country know when it is not only under attack but under attack by those so well resourced, or so uniquely focused, that it could not be other than a hostile state?[15] More specifically, what differentiates a thousand talented hackers working for some ministry of foreign intelligence and hackers of similar numbers and quality associated only through some hard-to-penetrate secret-handshake hacker-net?[16]

Mere proficiency may not be sufficiently telling – but might other attributes?[17] The ministry would be more secretive than the hacker-net, in the sense that if it finds something, it is unlikely to broadcast it. But given

---

[14] To be sure, "give me this" coercion is hard to do without identifying "me." But some terrorists do act as if wreaking anonymous violence for general ends may be worthwhile.

[15] The scenario assumes that the point of hacking is not to transfer resources to the attacker – because if it were, the first good clue to what is going on would be who is receiving the resources.

[16] Presumably, the difference would be subtler than the ministry's hackers sending their packets from sites with Chinese addresses while the hacker-net did not – but who knows? See Bradley Graham, "Hackers Attack via Chinese Web Sites," *Washington Post*, August 25, 2005, p. A1.

[17] Some cyber-sleuths have concluded that the difference between hackers-are-us and military hackers can be found in the latter's error-free, "methodical and voracious" techniques. Shawn Carpenter described what are believed to be hackers working for

its bent toward compartmentation, the ministry is also less apt to engage in the spirited communitywide back and forth that fosters innovation. It may scoop up good ideas but not trot out good ideas for feedback. It is unlikely to pose problems with enough details to get pertinent solutions back. Even today, more holes are found by hackers whose hats are white (those who seek and sometimes publicize flaws as an avocation) than those whose hats are black.[18] We know this because most novel hacker attacks have been found in the laboratory first. The ministry is also likely to gravitate toward attacks that are unique but have seen little real-world testing or discussion and are thus potentially brittle.[19] But finding a clever attack, as such, does not prove much.

Specificity may be a better indicator. The ministry is likely to concentrate on what it deems strategically important targets while the hacker-net would more likely scatter efforts according to each hacker's obsession with his or her bête noire. The ministry can also maintain an offensive over an extended length of time because it can throw people into round-the-clock coverage. A high ratio of damage per intrusion is also indicative of state effort. By contrast, hackers tend to get more excited by penetration than subsequent excavation.

The pattern of self-revelation may also be indicative. Here, the ministry has a great challenge. If its operations are hard to distinguish from what

China's military (an espionage ring that federal investigators code-named Titan Rain), thus:

They always made a silent escape, wiping their electronic fingerprints clear and leaving behind an almost undetectable beacon allowing them to re-enter the machine at will. An entire attack took 10 to 30 minutes. "Most hackers, if they actually get into a government network, get excited and make mistakes," says Carpenter. "Not these guys. They never hit a wrong key" (Nathan Thorburgh, "The Invasion of the Chinese Cyberspies (and the Man Who Tried to Stop Them," *Time*, September 5, 2005, p. 35).

[18] Eric Rescorla goes so far as to argue that holes found by white hat hackers should not be publicized even if the patch is offered as part of the dissemination process because black hat hackers will be able to develop an exploit for the hole before enough patches are actually installed. This argument is based on the presumption such holes were previously unknown – suggesting that black hat hackers do not find so many holes on their own. Eric Rescorla, "Is Finding Security Holes a Good Idea?" www.dtc.umn.edu/weis2004/rescorla.pdf, 2004.

[19] It is a well-accepted tenet among cryptography professionals that a coding algorithm whose parameters have been published, reviewed on the outside, and found secure is safer than an encoding mechanism whose algorithm is a secret and therefore has not been tested by outsiders.

hackers do anyway, then the victims are unlikely to raise their a prioris for filtering beyond what was already warranted by the increasingly polluted environment of cyberspace. Once some distinguishing characteristics are available (what the attackers are after, for example), then defenses go up, at least as necessary to protect what the ministry itself might target. Mass surprise is probably another good indicator of state activity.

All told, state sponsorship is more likely if the attack method is highly unique, continuous, precisely targeted, and intended to exploit weak spots in the victim's chain of operations rather than simply its security. But these are only probabilities. If attackers really want to hide themselves in the noise, unambiguous knowledge of "who did it" is not a luxury possessed by those who would dissuade them successfully.

11.2.2 Headless Horsemen

Our story continues:

The United States gets tough. First, it raises the intensity and sophistication of its law enforcement activities to bring the guilty to justice. But little results. Victims are reluctant to come forward, much less install real-time packet monitors on their systems. Other nations are less than enthusiastic about subjecting their citizens to aggressive U.S. cybercrime legal practices. Several key politicians abroad veto collective actions against data-laundering sites for fear of uncovering other washday activities. Frustrated, the United States deems the best defense to be a robust offense. Active penetration of the information systems of other nations' state security organizations commences. Information warfare support systems are disabled upon discovery. Retaliation in kind against likely perpetrators follows swiftly and without legal niceties. Broad operational authority is delegated to operatives under "tell me later" rules of engagement. But the United States is not alone in going on the offense – so have other nations comparably endowed with skilled hackers; so too have super-patriotic groups of all national persuasions. Source ascription for hacker attacks is not absolutely flawless, and so some retaliatory packet bombs are sent to the wrong address. Worse, there is no global consensus on observing a firebreak between acts of cyber violence and acts of physical violence. To some, attacks on systems that preserve life directly (by providing safety, warning, traffic control, and such) or indirectly (by providing weapons control, for example) cross the line. On more than one occasion, acts of retaliation in cyberspace have culminated in violence.

Hacking calls for a continuum of policy responses. A rogue hacker may be subject to arrest and prosecution. If the rogue hacker is responding to direction, then those up the chain of command may be sought. Past some point, though, governments may be held responsible for actions carried out by their employees, or perhaps even their citizens. Even if direction is not explicit, the state, it can be argued, should have known about and suppressed such activity.

The transition between seeing hostile conquest as crime and seeing it as war is hardly clear-cut. Approaching it as crime finesses the diplomatic issues – which may resolve themselves as regimes rethink their national security policies. Compared to war, prosecution is cheap; even the costs of very large legal cases – Starr versus Clinton, or Klein versus Microsoft – are less than that of a jet fighter. If done in conformance to international law, resort to the courts can enjoy broad worldwide support in ways that unilateral violence cannot. Although they are neither swift nor certain in their effects, they are an indisputable part of any government's response kit.

The decision to take the offensive should be predicated on what the offense/defense balance is – no easy thing to measure. The control of information about information warfare is itself an aspect of information warfare. Up to some point, a small amount of effort in computer attacks requires a much more expensive countereffort in defense to restore equilibrium. Indeed, so small is this effort that it can be supplied by schoolchildren (hence, "script kiddies") as easily as it can be by determined adversaries. To avoid embarrassment, if nothing else, requires defenses at some level, such as password protection, port closures, firewalls, configuration management, and the withdrawal of super-user status from many user-launched processes. The measures are particularly important given that most hacker attacks are already anticipated in theory but succeed because of breakdowns at the human and organizational level.

Beyond that point, the game gets harder. With known holes plugged, both sides would look for unknown holes so that the attacker, if first to find one, can exploit it or, if a defender finds it first, he can plug it. The side that discovers one has the initial advantage, but not necessarily a decisive advantage in a multilayer defense system. Unlike nuclear warfare, where the prospects of a permanently crippling first strike riveted attention, only fleeting advantages can accrue from a surprise computer network attack – unless used to facilitate physical violence and only if the violence is decisive. After the exploited hole is found and filled, the damage report will dwell on whether and when people can spot and override dumb computer decisions, how well insulated mission-critical operations are from everything else, and how fast systems administrators can resume trustworthy operations. The intensity curve in information warfare, which starts by favoring the attacker, levels off. Since a high-intensity attack terrain described here is largely unexplored, the shape of the offense/defense effort ratio at that stage is hard to know: does it take ten dollars or ten cents worth of defense to neutralize another dollar's worth of effort by the attacker?[20] Perhaps, beyond some point, no extra effort expended on offense can damage the defense further.

The offense/defense curves may twist over the course of combat as well. If hackers march out in full force early, the transition to the new curves may be swift. The intensive search for holes in operating systems and common applications would be matched by an equally desperate search for plugs (as well as the deletion of live vermin left behind). This may well exhaust the search space for holes[21] – perhaps even faster than they are increased every time a new operating system or another

---

[20] These days, defenders (including rich corporations) have more resources to spend than attackers (that is, hackers). In a war, when hackers work for militaries, both sides would start on a comparable footing. Thus relative efficacy will determine the net result of both sides pouring money into the effort.

[21] There are a few basic ways to look at holes in computer systems. One model is that there are only so many holes, so that a more intensive search by defenders and offenders will exhaust the possibilities sooner. A second, variant, model is that there is an indefinite number of holes but that the damage possible from exploiting the holes is either less consequential as the more troubling ones are found and patched, or only comes up under an increasingly rare set of initial conditions (such as if the user is in a particular mode and tries to take a specific action). A third perspective is that the easier holes are found first, leaving behind later holes that are just as destructive if exploited but are increasingly harder to find. (Both models assume that whatever fixes are made in response to discovered vulnerabilities actually work and make no new holes.) The difference between the two models is significant. If the first two are good characterizations of computer vulnerabilities, then the potential for information warfare *does* decline over time; by the same logic, open-source operating systems, where defenders have the advantage of more eyes having seen the code, and attackers have the advantage of getting to see the code, are also safer. If the latter is true, security will not necessarily improve over time, and while the bugs that *are* exploited will be more obscure, they may be just as damaging. As Ross Anderson argues, the theoretical case that open systems are more or less secure has yet to be convincingly made. Ross Anderson, "Security in Open versus Closed Systems: The Dance of Boltzmann, Coase, and Moore," www.ftp.cl.cam.ac.uk/ftp/users/rja14/toulouse.pdf, 2002.

core system product can introduce new ones. Product introductions may slow down, though, if customers, aware of the wartime hacking environment, demand that systems are bug-proofed before new versions enter the environment.

11.2.3 Perfect Prevention

And so the story moves on:

Whatever benefits the information revolution might bring, it was not worth creating a new casus belli for World War III. Could national security be secured by disarming the world of information warfare weapons? First the United States tried to leverage its companies' dominance of the global software market to restrict the export of hacker tools unilaterally. But most such tools were little but friendly user interfaces atop well-known techniques; they could be easily replicated overseas and passed via the Web. Attempts were made by the U.S. government to negotiate back-door agreements with major software vendors to have them write their software in ways that would help U.S. authorities trace hackers. But they were resisted by companies that viewed themselves as less national than multinational. The continuing rise of open-source operating systems and related software meant that, in some cases, there was no one with whom to negotiate. Next came international proposals to retard the release of software programs until they could pass tests that proved them to be hacker-proof – but no one could devise any such tests.

A good deal of what the government might do to improve computer security is not unique to governing. Yes, it can fund research into underlying technology, test methods, and standards development – but so could any sufficiently endowed institution. Although the government can set a high standard for computer protection by how well it guards its own systems, so should any institution that serves the public (such as a credit card company instituting a credit card authorization system).

In some cases, new laws may make a difference; they might hold software manufacturers more liable for the damage done by hackers because their software has holes.[22] Recent trends, however, have been otherwise.

---

[22] See Nancy Leveson, *Safeware*, Reading (Addison-Wesley), 1995; National Research Council, *Computers at Risk*, Washington (National Academy Press), 1991, especially the latter's argument about software liability.

The Digital Millennium Copyright Act can be used to criminalize publicity about defects in software.[23] The software industry tried to promote state passage of the Uniform Computer Information Transactions Act (UCITA), which tilts the burden of proof away from software companies.[24] Although the Health Insurance Portability and Accountability Act (HIPAA) of 1996 scowls at those who reveal private health records deliberately, it casts more forgiving eyes upon those who allow them to leak inadvertently.

In recent years, the insurance industry seems to have built up enough confidence in the statistics of cyber-crimes to offer insurance against some of more common problems, such as viruses, denial-of-service attacks, and leaks of intellectual property from intrusions. If this market develops,[25] something that has yet to happen, insurers may develop standards of due diligence, which, in turn, can become the basis for legal decisions about how much responsibility system owners bear for attacks on them.

Otherwise, most infrastructures being largely private,[26] their owners are not responsible to the federal government, and whatever "public service" mandates exist (as for utilities) tend to be imposed by local governments. Among the various sectors, federally insured banking and, for different reasons, defense contractors are unique in being subject

---

[23] In the summer of 2002, Kevin Finisterre, a consultant with the security firm Secure Network Operations, had let Hewlett Packard (HP) know of nearly twenty holes in its Tru64 operating system. But in late July, when HP was finishing work to patch the flaws, another employee of Finisterre's company publicly disclosed one of the vulnerabilities and showed how to exploit it – prompting the technology giant to threaten litigation under the Digital Millennium Copyright Act. Finisterre now says he'll think twice before voluntarily informing another company of any security holes he finds. See Robert Lemos, "New Laws Make Hacking a Black-and-White Choice," http://news.com.com/2009-1001_3-958129.html, September 23, 2002. HP dropped the suit.

[24] Maryland and Virginia passed UCITA, but the act then lost momentum and no other state passed similar laws.

[25] In late 2001, some analysts projected that $2.4 billion would be spent on such insurance by 2004. Sandra Swanson and George Hulme, "Insuring against Cybercrime Gets Tougher," *Information Week*, January 7, 2002, available at www.informationweek.com/story/IWK20020104S0022. However, the market has not developed as hoped.

[26] Although many governments own a higher percentage of their nation's infrastructures than the United States government does, the trend in most places is toward privatization.

to government-imposed due diligence strictures.[27] Others may have no good reason to cooperate with the federal government, and many firms have to be led kicking and screaming into cooperating with investigations of crimes against them.[28] Will cooperation win private companies privileged insight into new technology or indicators of warning? It seems unlikely, on two counts. The government has very little to offer in either category; nearly all such technology is already commercial, and almost all warnings of impending attacks are false alarms. Even if it had something to offer, the legality of sharing information with some but not all systems owners is dubious. Suppose that the government warns a cooperating Dominion Power but not an uncooperative Pepco of an impending attack. Suddenly, Pepco is hit with an attack; it may thus have grounds to complain loudly enough to make the government regret having played favorites.

11.2.4 Total Transparency

The result of these measures follows:

Henceforth, everything in cyberspace would be open and aboveboard. International conventions against cybercrime were not only crafted but also made enforceable through the widespread installation of packet sniffers, the mutual abjuration of encryption, and a code of investigative conduct. They were the twenty-first-century version of the Kellogg-Briand Pact. Militaries and security agencies returned their networks to their erstwhile air-gapped state, and international law presumed hackers to be personally and criminally liable irrespective of

---

[27] The Information Security Forum, an international security association, calculates that many of its members expect to spend more than $10 million on information security controls to comply with regulations laid down by the Sarbanes-Oxley Act. See "Report: Sarbanes-Oxley Could Threaten Security," http://news.com.com/2100-7348_3-5783472.html, July 11, 2005.

[28] Corporations may reasonably worry that giving government access to its information systems makes it easier for the government to find other compromising information. Nevertheless, over a ten-year period, the government has persuaded several industrial groups to form or join Industry Sharing and Analysis Centers (ISAC) with the purpose of exchanging information on security threats (which since 9/11 the government has broadened the definition from cyber-security to security against terrorists). Some ISACs have government participation and some do not. The Department of Homeland Security (DHS) now runs the program. See the DHS Web site at www.dhs.gov/dhspublic/display?theme=73&content=1375.

to whom they reported. But that hardly cured matters. First, these laws seemed to have little effect on private hackers or those in the pay of organized crime. Second, state security organizations were suspected of using third parties to do what they, themselves, were forbidden to do – further complicating command and control over such activities. So the screw turned further. A global public key infrastructure was established. The world's major ISPs, router operators, and IP-based telephone switches were forced to reject unsigned packets and forced to archive all encrypted communications. Some types of hacking vanished – but so did the ability to protect privacy from commercial firms (who could now readily trace packets back to real-world addresses) or government censors. No longer did people discuss legal, medical, or psychological issues on the Internet. In the end, it was not so hard to finagle people into giving up their access keys and associated PIN numbers, thereby nullifying the trustworthiness of their authentication and hence their security devices. No one knew this better than security agencies.

Since cyber-weapons are ubiquitous, the most straightforward way to reduce cyber warfare among states may be to make cyberspace more transparent.[29] But did the logic of deterrence – that states were more secure relying on mutual assured destruction than they were in competitively investing in weapon-defeating systems – hold in cyberspace? Can one state assure others of its good intentions by making itself so vulnerable that it falls as the first victim whenever mischief passes some threshold? Should it make its cyber-security mechanisms visible to other states? Should it make the health of its databases, servers, applications, or information services dependent on how comfortable others feel about their own infrastructure?[30]

That all depends on how foolish the state wishes to be.

First, many information security features are also safety features that ought not be disabled. The robust use of digital signatures protects not only against corruption from the outside, but also corruption from the inside, and it permits human or mechanical error to be traced back to its source. The same holds for fail-safe mechanisms that have humans

---

[29] There are less drastic approaches to global information assurance regimes. See, for instance, Lorenzo Valeri, "Securing Internet Society: Toward an International Regime for Information Assurance," *Studies in Conflict and Terrorism*, 23, 2000, pp. 129–46.

[30] To the extent that interdependence gives each state a stake in the assumed cooperation with other states, it may promote peace in much the same way that deeper trade relations do (although dense economic ties were not enough to keep the deeply intertwined Soviet republics from separating into marginal subeconomies).

review certain computer decisions. such as moving large quantities of money and erasing critical files.

Second, mutual access to many of the really interesting defense systems (such as nuclear command and control) may reveal secrets or corrupt operations with consequences more serious than any damage that can happen to cyberspace per se.

Third, one can falsely appear vulnerable. For instance, is the system portrayed as critical and hence opened to inspection one that another system really relies on – or is there a backup system that sits snugly behind some firewall and can be turned on if the first goes bad? At least as information systems grow more complex, the cost and confusion of maintaining fictions rise disproportionately.

Fourth, third-party mischief becomes too easy. Such hackers could penetrate defenses of one side, which then blames the other, and the resulting mutual recriminations disable the entire information system of both. True, systems permeable to each other need not be permeable to everyone. Yet, if each side is so confident in own security systems that it need not worry about third-party hackers, then both sides probably need not worry about each other either.

Besides, all this mutual assured destruction stuff was a product of a bilateral era. There is no "mutual" in America's unilateral moment. Furthermore, the United States is far more dependent on its own information infrastructure than is the case, for example, for North Korea, much less al Qaeda.

11.2.5 Nasty Neighborhoods

Our story culminates in security through invidious distinction:

After four successive tries, the U.S. government admitted that it could think of no good way to clean cyberspace other than to do what it was already doing: inveighing against the inevitable, enforcing laws as best it could, and keeping warfighters up to date on the newest techniques. But the story does not end here. The ever-rising din was bad for business if nothing else. Organizations did what they could to bulwark their own networks and cope with their increasing complexity by pouring time and money into prophylactics, intrusion detectors, testing, retesting, redesign, root cause analysis, and training. Yet they fell behind, in large part because they were able to make little headway against their own feckless employees, coupled with the fact that their customers, by definition, came from outside the firewalls. So business subsidized the establishment of digital signature registries and started to demand them of their business customers – extending this habit to their e-mail correspondents, and later to their ultimate consumers. Similar filters were established on e-mail systems, especially those held by ISPs. The Internet bifurcated into two types of transactions: those that were digitally signed and those unsigned – both, it should be added, which ran over the same infrastructure. There were safe neighborhoods and not-so-safe neighborhoods; the latter were for idle and salacious chatter, gambling, or exchanging warez (illicitly cracked computer programs). People who ventured into the latter were often defrauded or beset by strange computer diseases but they supposedly knew the risks they were taking. Legitimate businesses, by and large, stayed away or entered such terrain through cutouts [third parties]; they were, on the whole, safer than before, even from insider attack. Many people, unfortunately, found themselves on the wrong side of the tracks willy-nilly, much as so-called "combat zones" are rarely found in fancy suburbs. They could not get digital signatures because they had too little money, too much past, or too many unsound opinions. This hurt worse when telephony shifted to the Internet backbone and getting an education, health care, a car, or a house without Internet access became complicated. Few commercial firms wanted to deal with the great unsigned. With all the "good citizens" safer and sounder, everyone else was far more likely to be victimized because governments had other things to do.

The tension between private and public responsibilities for protecting cyberspace is already apparent. Will the Internet retain its unified architecture? If security concerns worsen, several infrastructures (including call control, electric power management, financial transfers, and air traffic control) may maintain a separate status with tightly guarded access.[31] To the extent that quality of service is an issue, specialized internets may arise to prioritize certain traffic by its intrinsic importance or to overcome hiccups in flow rates (for example, for synchronous audio or video) that would result in unacceptable service. Broadband networks might split

---

[31] Admittedly, this is an idea that comes and goes. For instance, *Business Week* that reported that IBM was betting that its nascent Global Network would quickly become "the businessperson's Internet." "IBM Swings Into The On-Line Fast Lane," *Business Week*, December 5, 1994, p. 100. A later *Business Week* report covered the announcement by MCI and BT that they "would offer what amounts to first-class service for businesses on the Internet – a separate parallel network with extra capacity, especially on international routes, and duplicates of many popular Web sites. The service is guaranteed to be more lightly trafficked so packets of information can be moved more quickly and dependably" "Limo Service for Cruising the Net," *Business Week*, June 24, 1996, p. 46.

off to couple content and service more effectively (for example, a cable owner may wish to give its own material priority status[32]). Raising the sensitive parts of the Internet beyond what the masses enjoy may increase the security of the lucky contingent[33] but decrease the security of what is left as powerful users lose interest in the latter, much as walled suburbs inure people to crime in the city. Finally, if Internet users are billed by packets sent, distributed denial-of-service attacks may decline once the owners of zombies (computers programmed by malevolent parties to spam other sites) start getting bills for their system's behavior[34] – which, conversely, may explain why such billing will never happen.

If the government cannot make system owners protect themselves, should it nevertheless be responsible for their protection? Should it provide insurance for those who insist on building in the flood zones of cyberspace? The government's hand-wringing about society's vulnerability to attacks in cyberspace may be signaling potential adversaries that it can be deterred through such means. After all, foreign strategic literature, notably China's, has a soft spot for the stratagem – a lightweight but precisely aimed stab at an adversary's soft spot that neatly and quickly resolves the matter. And what better soft spot than America's highly trumpeted dependence on information technology? Generate enough pain – and prove that it can be repeated – and the U.S. government may think twice about steaming out to intervene overseas.

---

[32] See, for instance, "Tolls Could Dot the Internet Highway," www.cnn.com, February 27, 2006; Declan McCullagh, "Republicans Defeat Net Neutrality Proposal," http://news.com.com/2100-1028_3-6058223.html, June 8, 2006.

[33] Unless the owners foolishly develop proprietary but, for that reason, poorly tested security protocols in the process.

[34] By way of comparison, security expert Steve Gibson found that his 3 megabit per second connection to the Web was simply flooded by a distributed denial-of-service (DDOS) attack involving fewer than five hundred zombies, a population largely recruited from cable modem owners with, if they are typical, always-on connections to the Internet and no firewalls in place. Brian Livingston, "Windows XP and DDOS," *Infoworld*, March 11, 2001, p. 60. Gibson further notes that Windows XP's ability to send out packets with fake IP addresses would defeat DDOS defenses that rely on painstakingly blocking service from specified zombies. Munir Kotadia reported that "Virus authors are choosing not to create global epidemics – infections of the type caused by Melissa and Blaster – because that distracts them from their core business of creating and selling zombie networks." Munir Kotadia, "Experts: Zombies Ousting Viruses," http://news.com.com/2102-7355_3-5720428.html, May 25, 2005.

The prospect is tempting but temptation is dangerous for all concerned. Perhaps instead, the U.S. government can credibly let friend and foe alike know that it cares little what befalls the feckless in cyberspace.[35] Hence there is no basis for coercion. As for attacks on government-owned systems, if the U.S. military cannot take any other nation's best cyber-shot, it has only itself to blame for choosing not to protect systems that could be protected. The same can be said for other critical public systems, including air traffic control, law enforcement, and social security. Beyond the public domain, system owners who let others muck with their information can be usefully and *correctly* blamed for whatever failures in the real world result from their errors in the cyber world. Let the public vent its wrath correctly. At very least, the public's anger should not be deflected by the government taking responsibility for systems over which it has no power.

To be sure, there are laws to be enforced, and states are on firm ground asking other states to help fight crime in a medium that respects no state boundaries. Still, cooperation with even friendly states[36] is not absolute in far worse criminal cases – and much of what is entailed in computer hacking has only recently been made a crime.[37] Such limitations suggest

---

[35] Hurricanes (Katrina in 2005, $100 billion in damage; Andrew in 1992, $25 billion), earthquakes (Northridge in 1994, $15 billion), snowstorms (the Blizzard of 1996, $10 billion), floods (the 2002 floods that hit Central Europe, also roughly $10 billion), and droughts cause great damage with little evidence of their passing in GNP statistics. How much damage, by comparison, can information warfare cause? Michael Erbschloe of Computer Economics estimated the I-Love-You virus of May 4, 2000, by to have cost the world economy $8.7 billion. Hal Berghel, "The Code Red Worm," *CACM*, December 2001, p. 19. The virus supposedly affected 12 million users at its peak (yet, subsequent surveys suggested only one in fifteen U.S. companies suffered substantial disruption; see Evan Hansen, "Poll Finds Few Affected by 'I Love You' Virus," http://news.com/Poll+finds+few+affected+by+I+Love+You+virus/2100-1023_3-241539.html, June 6, 2000). But that assumes that the lost time was worth $800 per person despite the fact that only a very small fraction of these people are paid so much or were rendered so useless as to be sent home. A closer estimate would probably be at least one order of magnitude lower.

[36] States can also sign treaties that prohibit certain forms of information warfare. Insisting on clauses that would facilitate law enforcement investigations directed at hackers may help. As with many such treaties, enforcement is problematic, especially when detection is so hit and miss.

[37] It took until 2001 before Europe passed a convention on cybercrime (for a copy, see conventions.coe.int/Treaty/en/Treaties/Html/185.htm).

why strategic information warfare – should it, in fact, come to pass – may disenfranchise the state as much as the specter of nuclear warfare empowered it.[38]

If the privatization of security in cyberspace makes sense, why not encourage vigilantes in cyberspace? Suppose that the victim nation fingers the likely suspects, or they reveal themselves. It then slips to talented mischief makers a few under the table hints about the adversary's weak spots and presto: the fogs of war have been loosed in ways that prevent counterdeterrence, since the hackers cannot be recalled. But, then again, this may drag everyone into a war whose echoes die out all too slowly. As it is, difficulties in controlling the level of cyber-violence mitigate against liberating the super-patriot hackers – as if it mattered. They are unlikely to ask for permission if aroused. But they need not be fed juicy little tidbits on adversary vulnerabilities, either.

11.3 Exploiting Unwarranted Influence

Nations can exercise influence over cyberspace in ways that precede the methods of friendly conquest discussed in the preceding sections. Invention, for instance, counts for something; the fact that telephony was invented in the United States has won it the shortest country code ("1"); ditto for the Internet and the fact that most U.S. addresses dispense with a ".us" suffix. Pace Joseph Nye, nations can exert soft power or cultural influence without exerting hard power. Cyberspace, indeed, has been very very good for the promulgation of U.S. culture, and, to some extent, U.S. values. The power of prominent U.S.-based firms has lent the nation no small influence in setting rules and de facto standards. Yet, these are instruments of the nation, not the state.

To reiterate the mechanism described in Chapter 6, one way of exercising power would be for the would-be conqueror to give away or sell for a pittance cyberspace assets such as information, services, or connections. This need not be expensive; although creation is hard, duplication

---

[38] Nuclear warfare empowered states by permitting them to use the threat of direct consequences to put national security in front of more everyday economic or political concerns.

is easy, as long as little customization is required of the customer. Users of cyberspace begin to rely on what they get. Their desire to produce alternatives to such information, services, or connections fades and their ability to do so may then wither; meanwhile, their desire to produce binding complements rises. Over time they adapt their own regions of cyberspace to foster compatibility with the regions of cyberspace into which they are offered entry. They may grow so dependent on such access that they would pay for what they got for free, whether in coin, quid pro quo, or simply continued commitment to their relationship.

Those who wish to gain influence for their version of cyberspace might keep two other factors in mind.

One is network effects. As Brian Arthur and others have observed, small, even random, initial advantages of the sort that may be inferred by discussions in Chapters 6 and 7 can often widen into a lead too strong to be easily overcome. Take information technology standards.[39] Typically, one firm has a proprietary interest in one standard, and its rival in another. Standards compete by attracting users, third-party applications (such as those that run atop operating systems), conforming platforms (those that run beneath operating systems), and support service and training providers – a network or ecosystem, as it were. Initially, people choose the one that best fits their individual needs. Large organizations may strike deals with one or another vendor to adapt or adopt it to their own requirements. At some point, however, the uncommitted have to choose one standard or another. Features matter, but the standard that has attracted the better, which often means larger, network is often preferred for that reason alone. With more vendors supporting the standard, users usually get more choices for their money. If the standard affects communications, they can often reach more of their counterparts with it. Even if it is the prospective market share that matters, the best predictor of that is usually the current market share. So, bandwagons form, with the winner systematically overtaking the loser.[40] Just ask Sony, whose

---

[39] For background, see a seminal article on the topic by Charles Morris and Charles Ferguson, "How Architecture Wins Technology Wars," *Harvard Business Review*, March–April 1993, pp. 86–96.

[40] The overall structural outcome need not necessarily be to anyone's liking. Tom Schelling demonstrated as much by creating a heuristic model that started with a grid in which

technically superior Beta format was orphaned as nervous users realized that they would soon have no material to play on their machines if and when VHS became the preferred medium.[41] With a race so close that a push in one direction is decisive, the winner need not overpower the loser but only gain an edge at the right time – for example, when the technology's maturity or the emergence of cofactors indicates that it is breaking into new markets.

With most industries being information industries to some extent, such phenomena could become widespread. Nevertheless, cyberspace is particularly prone to such lock-in. The marginal cost for producing the next copy of information (such as software or access to Web sites) is close to zero, and thus almost all revenue from the next customer goes straight to the bottom line and thence to creating greater attractiveness. Proliferation rates in cyberspace are accelerated by the fact that information can be distributed so quickly and so widely; interaction rates are also accelerated there because interaction is easy. Conversely, cyberspace is good at agglomerating scattered users around the world; such self-defined communities can hold off trends better than if individuals were isolated.

This leads directly to the second factor: the importance of pricing at marginal costs, at least initially, rather than for expeditious investment recovery. This matters when the broader goal is to build dependence for the sake of structuring future relationships and the system opened up for enticement is one that would have existed in its present, albeit closed, form anyhow.[42] If small differences in attractiveness lead to large differences in adoption, it would be a shame if the mandate to recover

each house has eight neighbors. People moved to random free locations whenever they had fewer than two neighbors of their own race. The result was rigid segregation – a result no one intended. Tom Schelling, "On the Ecology of Micromotives," *Public Interest*, Fall 1971, p. 25.

[41] To be fair, the cost of achieving higher quality was a tape that was not long enough to record full-length movies.

[42] This rule does not necessarily apply to systems that have to be built from scratch or those for which the market is completely uncertain; here a mandate to recover costs may be a salutary prophylactic against pouring money into one half-baked idea after another.

costs led to prices that dissuaded early adopters, the paucity of whom doomed attempts at general adoption in the face of competition.

Creating a cyberspace that fosters dependence requires either that the conquered fails to foresee adequately where the sequence leads, or sees it but calculates that the benefits are worth the dependence when all the sums are made. Whether dependence, however, *will* be accepted may rest on broader factors. Take GPS. The signals are free. Reception equipment now costs less than $100. The signal itself contains virtually no semantic standards of the sort that would alter how geospatial information is organized or conceptualized. For almost all civilian and many military users, it is a completely new capability that does not replace an existing competitor with sunk capital who could raise a political fuss. It is the closest that DoD, which did not make its reputation as a charitable organization, has ever come to giving the world a gift with no strings attached. Yet, somehow the Europeans are not satisfied with this deal. In March 2002, the European Commission voted to invest nearly a billion dollars in full-scale engineering of an alternative constellation, Galileo. By 2005, the European Union had made its first demonstration/validation launch. Galileo's proponents cite its superior or at least additional technical characteristics, and, understandably, multiple signals do enhance accuracy and availability of location information. Europeans also express fears that GPS could be pulled out of service at the whim of the U.S. government. The true argument, made sotto voce, may be that the Galileo program gives European aerospace firms a shot at improving their electronics manufacturing expertise at taxpayer expense. The U.S. aerospace industry, while the world's best, is not so far ahead that European firms cannot pursue it in the marketplace. Were the U.S. lead insurmountable, maybe the struggle to fight off free service would not be as strong. Be that as it may, friendly conquest is not trivial.

The controversy over the Internet Corporation for Assigned Names and Numbers (ICANN) shows a similar dynamic at work. ICANN was chartered by the U.S. government to internationalize the management of the Internet, specifically the assignment of domain names and numbers. ICANN strives toward inclusiveness in its management structure and holds board meetings all over the world. Yet, governments, especially in the third world, keep trying to replace what they see as a tool of the U.S. government, however exaggerated, with some form of UN governance.[43] Little of this has anything to do with ICANN's decisions. Indeed, a prominent 2005 controversy over the creation of a ".xxx" domain for pornographic material pitted ICANN against the U.S. government.

A U.S. government strategy of friendly conquest must also overcome no small *internal* resistance. What should be opened for access, for instance? The more valuable pieces of information that the government possesses are either highly classified or are otherwise deemed sensitive. Bureaucracies like hoarding information, and not only because it lets them tell themselves, "I know something you don't." It maximizes their sense of control, especially over uncomfortable news stories. Revelation may also compromise their sources and methods as well as the promises they make to assure others of their anonymity.

The techniques of sharing are also problematic. Simply opening one's system to others may let them rummage through the rest of the space. Even if the information released into the common sandbox is scrubbed clean of sensitive material, giving others access to the sandbox may compromise the systems that pass it forward. Placing it outside the walls mitigates this danger, but unless information is transferred by hand or something of equally paleolithic security, interactions between the core enterprise infrastructure and this playground may provide a back door to the golden cellar. There are grounds for skepticism over the ability to secure these transactions. Bureaucracies uncomfortable with information sharing have no incentive to give or gather assurances on this account. Releasing information in pursuit of broader strategic goals that redound to the benefit of higher executive levels can be a tough sell. And grudging acceptance will not do; political bosses over, much less outside, such agencies may not know very well what is *not* shared and so cannot ask for it.

---

[43] See, for instance, Declan McCullagh, "U.S. to Retain Control of Internet Domain Names," http://news.com.com/2102-1028_3-5770937.html, June 30, 2005; idem, "Will the U.N. Run the Internet?" http://news.com.com/2102-1071_3-5780157.html, July 22, 2005; idem, "Bush Administration Objects to .xxx Domains," http://news.com.com/2100-1028_3-5833764.html, August 15, 2005.

Ingratitude is another concern. Everyone, for instance, seems to want the good stuff (for example, the information of greatest relevance to the crisis du jour), which tends to be what is most sensitive. Unless expectations are carefully managed, the never-so-grateful-as-they-should-be recipients will be disappointed with every act of known withholding. The greater their disappointment in not getting it, the less likely they would yield anything in return for such information themselves, and the less likely they are to offer up information they have of value.

The last difficulty is standards. Those with material to give out would probably like to hand it out in ways that heighten the recipients' dependence on them for updates and upgrades while sweetening the recipients' desire to adapt their own systems to consume their material intelligently. So, the recipients, not themselves fond of dependence, are going to demand open standards. These demands will be hard to deny in light of the fact that everyone, notably every government, pays lip service to open standards. Yet, adopting international standards for data formats and open APIs for procedures tends to reduce their dependence on input from the U.S. government. If, however, their own information is formatted using open standards rather than whatever proprietary standards they must adopt for compatibility with the data they get from the U.S. government, they may be able to get third-party software support from anyone. They do not have to rely on government-cleared vendors, apart from those who have developed niche products for federal customers.

Indeed, the U.S. government has been a constant advocate of standards to avoid its own dependence on software vendors and to ease information sharing among its own agencies. As it is, the days when governments were leading-edge consumers and manipulators of information technology are long past. For the most part, the government depends on armies of contractors to make even its own systems run. Man for man, it cannot compete with Microsoft.

11.4 Against Unwarranted Influence

Is the exercise of unwarranted influence in cyberspace a proper public policy concern? Maybe not; far more laws, properly, restrict hostile conquest (such as violent coercion) in the real world than restrict friendly coercion (such as love). Yet, some "voluntary" transactions in cyberspace are restricted: gambling, child pornography, extracting personal information from minors, and mishandling personal information from the rest of us. Fraud and blackmail in cyberspace are treated like fraud and blackmail in real space. Monopoly in restraint of trade is illegal. The Digital Millennium Copyright Act also restricts the use of cyberspace to infringe on intellectual property rights. Europe has declared that some of the coins exchanged in cyberspace (such as personal information) may be declared illegal tender or at the very least their exchange cannot be compelled. Nevertheless, it is not easy to legislate cleanly, much less wisely, the more subtle issues of standards, the patterns of dependence, or the exercise of unwarranted influence – antitrust laws aside.

11.4.1 In Microsoft's Shadow

This brings us to the institution that, for good or ill, has greater influence in cyberspace than anyone.[44] More decisions that affect the ability to carry out hostile conquest in cyberspace are made in Washington State than Washington, D.C.[45] Yet, no company has more aggressively and successfully shown how to achieve unwarranted influence in cyberspace through the manipulation of software and service offerings than Microsoft. It has been argued, for instance, that Microsoft's image of what an office should be – one dominated by the benevolence of management

---

[44] Circa the summer of 2005, Google's aggressive hiring campaign coupled with the introduction of products on its Web site persuaded Gary Rivlin to suggest, "Relax: Bill Gates; It's Google's Turn as the Villain," *New York Times*, August 24, 2005, p. A1. Joe Kraus, who founded the search firm Excite, observed, "In the 1990s, IBM was widely perceived in Silicon Valley as a "gentle giant" that was easy to partner with while Microsoft was perceived as an 'extraordinarily fearsome competitive company' . . . [but today] Microsoft is becoming IBM and Google is becoming Microsoft."

[45] In December 2001, Microsoft Chief Security Officer Howard Schmidt was picked to be the vice chair of the President's Critical Infrastructure Protection Board, and in that role was asked to help design strategies to protect U.S. computer systems from hacker attacks by al Qaeda or other terrorist groups. It is unclear whether leaving the state of Washington for the city of Washington increased or decreased his overall influence on the security of cyberspace.

information system folks – combined with its software monopolies has resulted in too few applications that feature true collaboration, the integration of multiple sensory inputs (such as voice and gestures), or robust version management.[46]

Microsoft has been persistent and clever in successively exploiting dominant market positions in one field (such as client operating systems) to enhance its ability to dominate others (such as office automation applications, and, more controversially, Web browsers). It has a very large market share for server-side operating systems (that is, Windows NT and its descendants), database servers, and Web servers, but it still faces competition in all three of them. Microsoft holds a smaller but growing share of the operating system market for personal digital assistants, game players, and cell phones.

Can Microsoft leverage its influence in software packages to wrest a position in cyberspace? Its early-1990s foray into set-top boxes that would control the "information superhighway" was halted once it became clear that the information revolution was not going to be televised but Webcast. Current efforts to gain the set-top box market are being stymied by powerful media companies that have grown wary of Microsoft's influence.[47] Similar distrust has hobbled its ability to extend its Pocket PC operating system from PDAs to cell phones.[48] Although Microsoft's Web sites in aggregate (including its free Hotmail service) get more visits than any other company's, converting eyeballs into consumer outlays has proven difficult. MSN, its Internet portal, continues to run a very distant second to AOL, despite its being one click away on most consumer Windows screens. Microsoft tried but failed to introduce Smart

---

[46] See Andy Oram, "Thinking Outside the Outbox," http://www.oreillynet.com/pub/a/network/2000/06/01/outbox.html, June 1, 2000.

[47] See "Microsoft Misfires," *Business Week,* December 18, 2000, p. 190.

[48] Sony's President Kunitake Ando commented, "We know that Microsoft is a big threat. If everybody goes toward Windows and Microsoft technology, it's not so good." "Ganging up to Compete with Microsoft," *Business Week,* December 3, 2001, p. 70. A subsequent *Business Week* article report: "Mobile-phone makers . . . have watched the software giant gobble up most of the profits in PCs and fear the same result in handsets. Says one telecommunications executive: 'Everybody hates Microsoft.'" "Will Microsoft Overplay Its Wireless Hand?" *Business Week,* March 11, 2002, p. 48.

Tags, "dynamic links" that converted keywords in documents created by Microsoft Word into Web links.[49] Such links would have, by default, pointed to Microsoft-favored sites. Vigorous protests forced Microsoft to withdraw that feature. Microsoft has also retreated from many high-profile experiments with its own content including the Sidewalk chain of city guides, the Expedia travel site (which it sold), and sites such as MoneyCentral.[50]

Microsoft's next thrust was to make itself simultaneously the marketplace for a new cornucopia of Web services (in hopes of collecting small change for every transaction) as well as the vendor for the software that would make it all this possible. Its ".Net" initiative would permit associated service providers to respond automatically to a Microsoft-standardized series of prompts. Its electronic wallet initiative, for instance, would permit people to access various Web sites by logging into a Passport account and having Microsoft in turn transfer this login information, together with associated data such as credit card number, to various Web merchants (see the vignette in section 8.3). Making such a complex system work in toto would require that all or nearly all of someone's applications *and business partners* belong to or at least recognize such a system. At a minimum, this would mean buying the appropriate application software

---

[49] As Michael Kanellos reported:

> For example, if a person was reading a story about traveling, the world "airline" could include a link that would divert the reader to an airline or travel service chosen by Microsoft. . . . Critics accused the company of reverting to old tactics by loading Windows XP with features such as Smart Tags, which give Microsoft some greater control over consumers' Internet use. . . . What was most worrisome for analysts and others is that Smart Tags tie Web content exclusively to Microsoft software, in this case Office XP and Windows XP, according to [Chris] Le Tocq, an analyst with Guernsey Research. The feature gave Microsoft "some powerful leverage,' Le Tocq said, particularly since the company can use its products to redirect users to MSN Web properties and eventually sites with "premium paid services". The test version included Smart Tags for sports, stock and university information (Michael Kanellos, "Microsoft Gets Diplomatic in China," http://news.com.com/2100-1001-932927.html, June 5, 2002).

> In early 2005, Google introduced AutoLink, a feature that would do similar things – the difference being that Smart Tags was built into the browser whereas users have to download AutoLink deliberately. Paul Boutin, "When Good Search Engines Go Bad," http://www.slate.com/id/2114308/, March 3, 2005.

[50] Jim Hu, "Competition: Ultimate Challenge to AOL," http://news.com.com/2009-1001-273801.html, October 19, 2001.

to create and respond to such messages. Shuttling these messages to and fro would necessitate, for all practical purposes, that they be plugged into the relevant switchboard – owned by guess who.[51] Sun Microsystems has developed its competing system, Liberty Alliance, and an accompanying set of services. As earlier chapters suggested, Microsoft's thrust in this direction has lost some power. Google, on the other hand, with its commerical success and scores of newly hired whiz kids from up and down the West Coast, may make a run for the goal, especially if its moves downmarket to browser software or even core operating systems.

11.4.2 Microsoft and Computer Security

It is scant exaggeration to note that the vulnerability of the global information infrastructure to various vermin rests, in large part, upon architectural decisions embodied in Microsoft's suite of software products. The Melissa and I-Love-You virus played off a feature in Microsoft Outlook that allowed e-mail to be sent to everyone who had ever e-mailed the user. In 2001, a doublet of worms (including Code Red and Nimda) exploited flaws in Microsoft's Internet Information Services (IIS) software; several hundred thousand servers were affected.[52] Entire programs such as Back

---

[51] Since Microsoft ".Net" standards are supposed to be available to anyone, why couldn't someone else compete to become the switchboard? Third-party software could, in theory, receive and transmit standard messages. But does anyone really believe that this operator could get away with running such a system without Microsoft server software, which is likely to be optimized to the next version of the ".Net" standard? And does anyone really believe that with all the defaults pointing to Microsoft's switchboard, and the only critical mass of Web server providers being plugged into such a switchboard, there would be any reason for a customer to go with someone else? More basically, as the *Economist* notes:

Microsoft has falsely portrayed itself as the champion of open standards in the past, notably during its "browser war" with Netscape, only to revert to its old tactics later. Might the company not simply be waiting for XML, SOAP and other new standards to take off, ask its critics, before hijacking them by creating its own proprietary versions? ("A Kinder Gentler Gorilla?" *Economist*, April 26, 2001, p. 63).

[52] According to security expert Steve Gibson, by contrast, the "last serious remote-code execution vulnerability to hit the Apache Web server was back in 1997. But IIS has them monthly." Brian Livingston, "Is It Secure? I Mean IIS" *Infoworld*, January 21, 2002, p. 39. Lest one argue this to be another case of market leaders getting a disproportionate share of attacks, Apache actually holds a comparable market share with Microsoft.

Orifice exist to exploit flaws in Microsoft Office. Much of the ability to penetrate enterprise networks and gain super-user access to files depends on the decisions embedded in Windows NT. As Peter Neumann had observed:

A well-known (but certainly not the only) illustration of these risk factors is Windows NT 5.0. It reportedly will have 48 million lines of source code *in the kernel alone*, plus 7.5 million lines of associated test code. Unfortunately, the code on which security, reliability, and survivability of system applications depend is essentially all 48 million lines *plus* application code.[53]

Microsoft's defenders argue that it is a target because vermin creators (1) know that viability depends on its spreading lustily from one to another machine, and (2) no other operating system has the critical mass to ignite a self-sustaining chain reaction. Apple computers, for instance, now account for no more than roughly one-tenth of all new machines and Linux has little desktop presence – thus malware specific to their platforms alone are unlikely to spread rapidly. Yet, for worms that operate against Web servers, the critical mass argument is weak: UNIX and Linux still account for a large share of the server market. Microsoft e-mail clients are also not as ubiquitous, but they host more than their share of vermin. Conversely, picking on less popular software has some advantages – there are fewer security experts and products in that market available to play defense. The claim that Microsoft products get picked on only because they dominate the market should, therefore, not be swallowed whole.[54]

Experts differ over whether Microsoft products are per se less secure than those of its rivals are.[55] It is probably not the best such product;

---

[53] Peter Neumann, "Robust Open-Source Software," 104, *CACM*, 42, 2, February 1999, p. 104.

[54] Several prominent security professionals have argued that it is Microsoft's monopoly position per se that poses a risk to security (irrespective of how conscientious Microsoft itself might be about making its own products secure). See Charles Greer et al., "CyberInsecurity: The Cost of Monopoly," http://www.totse.com/en/technology/computer_technology/cyberinsecurit171812.html, September 24, 2003.

[55] It does not help that "Three-fourths of computer software security experts at major companies surveyed by Forrester Research do not think Microsoft's products are secure." Reuters, March 31, 2003, 8:46 p.m., cited in http://news.com.com/2100-1002-994878.html.

BSD UNIX, for example, has a reputation of never having been hacked.[56] Fans of Linux make a strong case that with all the eyes[57] looking at the code of the popular open-source software, "all bugs are shallow ones."[58] Some security experts regularly pan Microsoft products,[59] but others rise to its defense.[60] A comparison of particulars does not leave Microsoft looking good. ActiveX, for instance, is far less secure than Java. Active Directories can create severe side-effects for the unwary. Hackers exposed the copy-protection mechanisms of Windows XP and thwarted the security mechanisms of Passport before either had hit the market.

Decisions on product design that weaken security may be influenced by Microsoft's strategy of ensuring that each Microsoft-supplied component is designed to work seamlessly with one another. The tighter the integration, the more a product in one niche can beat what a competitor offers because it works better with all the other Microsoft products already on everyone's desk. By now, Microsoft has no serious competition in office automation – a status it dearly wishes to keep. And so, e-mail clients, Web scripts, application macros, and applications themselves interact freely despite their each having different security requirements and expectations.

---

[56] The OpenBSD team brags that its operating system's default security has never been breached to allow remote privileged access to an OpenBSD server. Tom Yager, *Infoworld*, November 5, 2001, p. 58. Stephan Somogyi and Bruce Schneier write:

> One example of doing it right is the OpenBSD project, whose developers have audited its kernel source code since the mid-1990s, and have discovered numerous vulnerabilities, such as buffer overflows, before they were exploited. Such proactive manual scrutiny of code is labor intensive and requires great attention to detail, but its efficacy is irrefutable. OpenBSD's security track record – no remotely exploitable vulnerabilities found in the past four years – speaks for itself" (Stephan Somogyi and Bruce Schneier, "Inside Risks: The Perils of Port 80" 136 *CACM*, 44, 10, October 2001, p. 136).

[57] See, for instance, Joris Evans, "Developers Fast to Fix Open-Source Bugs," http://news.com.com/2102-1002_3-6057669.html, April 4, 2006.

[58] As famously argued in Eric Raymond, "The Cathedral and the Bazaar," http://www.tuxedo.org/~esr/writings/cathedral-bazaar/cathedral-bazaar/index.html#catbmain.

[59] Bruce Schneier, *Secrets and Lies: Digital Security in a Networked World*, New York (John Wiley), 2000.

[60] For example, Stuart McClure, Joel Schambray, and George Kurtz, *Hacking Exposed*, Berkeley (Osborne), 1999.

Microsoft products have many of the controls that permit users to set security at one of a variety of levels. Nevertheless, its traditional products required users to invoke them deliberately in order to achieve security and thereby sacrifice functionality in the process.[61] Administrators who are overworked and beset by user pressures for functionality often fail to make the right trade-off between security and ease of use.

In early 2002, Microsoft saw the light on security, it said, and launched a high-profile campaign to eradicate security-compromising bugs in its programs, thereby aiming to make its software "trustworthy." At a cost of several hundred million dollars, software engineers were pulled off their projects to participate in intensive security sensitivity sessions.[62] It pulled enough people into putting out a major patch on Windows XP (Service Pack 2) to delay its successor, Vista, until 2007. The corporation also swears that the products that are expected to follow its public commitment to security will default to safety rather than to promiscuity. At least for the first few years after its initiative was announced, the most obvious result of Microsoft's newfound devotion to security had been the near-daily announcements of serious software bugs in its own products. But mourn not Microsoft's woes. After all, it is always the *next* version of the software that fixes the bugs discovered in the last version – so security becomes yet one more reason for continual updates notwithstanding the risk of new versions introducing new security-related bugs. Nevertheless, the more that the quality of Microsoft products is linked to the national, nay global, well-being in cyberspace, the stronger the argument that its products have assumed the mantle of a global information utility and hence one deserving of regulation.

Nevertheless, suspicions that Microsoft has a hidden aim other than its own profits strain belief. Microsoft, especially among U.S.-based

---

[61] Under Microsoft's "Secure by Default" initiative, products are shipped with the security settings at their maximum level, which, in the nature of things, means that some applications that assume easy access everywhere do not run very well. Interview with Scott Charney, Microsoft's chief security strategist, in http://news.com.com/2008-1982-948386.html, August 6, 2002. Similarly, the highest setting of many spam filters (such as Earthlink's) accepts e-mail only from a designated list of friends – and thereby severely restricts functionality.

[62] Craig Mundie, Pierre de Vries, Peter Haynes, and Matt Corwine, "Trustworthy Computing, White Paper," paper prepared for submission to the Thirty-First World Economic Forum, available at www.microsoft.com/prespass/exec/crag/01-31trustworthywp.asp.

multinationals, is as patriotic as any and, in some ways, more than most.[63] It is a huge net exporter[64] at a time when the U.S. economy runs an enormous trade deficit. Their people populate many U.S. government advisory boards. They cooperate with the NSA and other intelligence agencies.[65] Yet, should the United States government conclude that the national interest is best defended by exercising influence in cyberspace, then it, more than any government, is in a position to make sure that Microsoft does not stand in its way. Conversely, Microsoft's dominance may make it difficult for any other government to exercise such influence.[66] And for Americans, at least, who applaud capitalism, distrust governments (especially other governments), and believe that two masters cannot conquer cyberspace, perhaps Microsoft's nonideological influence is a good thing.

11.5 Conclusions

Far too much ink has been shed on the subject of information warfare with little to show for it but the bureaucratic designs of the hungry and the fitful sleep of the innocent. This is not to say that everything is peace and light in cyberspace. Year by year it seems to become a more polluted

---

[63] According to Mark Montgomery, speaking on "Cyber Threats: Developing a National Strategy for Defending Our Cyberspace," before the Harvard command-and-control class of Professor Tony Oettinger on April 13, 2000: "Actually at the NSC we're pretty decent friends with Mr. Gates. We visit him and talk with his security people a lot. We need them. A lot of our military systems and most of our private sector systems are driven by his software. He certainly doesn't do our bidding, though."

[64] Roughly 60 percent of Microsoft's sales but less than a quarter of its costs comes from overseas. By contrast, most U.S.-based multinationals of Microsoft's size produce and sell around the globe in comparable percentages.

[65] One U.S. commentator has even suggested, perhaps implausibly, that the Microsoft's cooperation in facilitating the surveillance of Internet traffic may have been the price for its dropping of the antitrust case. David Winer, "Microsoft: Too Much Control of the Web?" news.zdnet.com/2100-9595_22-531040.html, November 8, 2001. Even the German intelligence services have expressed nervousness on this issue.

[66] This is not to say that other governments have not tried. The European Union has won suit against Microsoft in its own courts over the company's bundling of Media Player into its operating system. Even the Korea Fair Trade Commission has fined Microsoft $32 million (33 billion won) for abusing its market dominant position and ordered the software giant to modify the way it packages its Windows products. Aaron Tan, "South Korea Fines Microsoft $32 Million," http://news.com.com/2100-1047_3-5985332.html, December 7, 2005.

environment, forcing the rest of us to raise our doorsteps a little higher to keep the muck out. But raise it we do. Hence, while our shoes get dirty and travel is not completely reliable, our heads are largely intact. Warfare on information systems is not yet warfare on information.

Far too little thought has been given to the possibilities, for better or worse, of power in cyberspace that can be realized with seduction. Cyberspace is the commons in which tangible value can be transferred for the intangibles of relationships; at least initially, it can be for free. Like love, cyberspace may be a force for greater good or the more insidious forms of exploitation.

One can avoid exposure to others in cyberspace at all costs, abjuring networks or surrounding oneself with firewalls, never venturing beyond the familiar confines. But it is out there. Perhaps in their own way the two Bills had the zeitgeist right. Whether it was the "engage and enlarge" policy of Bill Clinton's (foreign) policy or the "embrace and extend" of Bill Gates's (standards) policy, the object was the same. There was and is a world to conquer.

APPENDIX A

Why Cyberspace Is Likely to Gain Consequence

Does cyberspace matter? Can societies deprived of their ability to use the commons of cyberspace (such as the routing infrastructure and standard software) nevertheless thrive unbowed?

It may be silly to conjure images of the horrors that would attend seeing the United States suddenly being thrown back to the primitive conditions of 1995, or, for most of the rest of the world, 2005. Yet, any topic that deals with war must be accorded at least a moment of seriousness. To ask whether cyberspace is the new high ground is just the latest version the age-old question: which medium dominates war? For example, did Britain's control of the seas prove decisive in World War I against a continental Germany?[1] Does air superiority win ground campaigns?[2]

The best answer to corresponding questions about cyberspace is neither "yes" nor "no" but "more so every day." This appendix lays out some key vectors of technology to illustrate both the changing nature of cyberspace and its increasing pervasiveness in the real world. Understandably, the tendency to extrapolate infinitely from the bright past has gotten many people into financial trouble, especially those invested in dot-coms. Yet, while stock markets rise and fall, and consumer tastes can be fickle, technology is well nigh inexorable.

Even in the highly unlikely event that microprocessors never get faster (and they have not gotten much faster since 2002), the eventual

---

[1] Some say no, e.g. Niall Ferguson, *The Pity of War*, London (Penguin), 1998.
[2] Some opinions from this author: no in Vietnam, contentiously yes in Desert Storm, unquestionably yes in Kosovo, and more likely than not in Afghanistan and the initial phase of the 2003 Iraq war.

retirement of all slower microprocessors in favor of new ones will raise the average speed of the entire stock of computers. Whatever security advantages come from using mainframe and minicomputer operating systems will shrink as applications once performed by such computers are replaced by those hosted on PC- and UNIX/Linux-based systems. The computer-savviness that characterizes a nation's youth cohort inevitably characterizes the nation as a whole.

That noted, this appendix examines several broad trends in cyberspace and, where apropos, limns their implications for friendly and hostile conquest there:

- More powerful hardware and thus more complex software
- Cyberspace in more places
- Fuzzier borders between systems
- Greater systems integration
- Accepted cryptography
- Privatized trust
- The possible substitution of artificial for natural intelligence

Other expected changes in cyberspace – more wireless presence, the proliferation of sensors, and the march of systems integration – can be found in chapter sections 6.4.1, 8.2, and 6.3, respectively.

A.1 More Powerful Hardware and Thus More Complex Software

Moore's Law, although expressed in terms of transistors per chip, predicts, for all practical purposes, that processor speeds double every two years. But for how long?[3] Feature size is the best clue to speed trends. In the late 1980s, when typical feature sizes were 1.0 microns, the theoretical limit seemed to be .25 microns. This would have suggested an imminent deceleration in improvement, but note that in 2005, companies began to

---

[3] "Moore's Law will prevail for at least 10 years, IBM researchers predict." Michael Kanellos, "Sam Palmisano's Technology Forecast," http://news.com.com/2100-1008_3-6054924.html, March 28, 2006.

experiment with "immersion" lithography to manufacture chips using .065 micron features – well past the older theoretical limit.[4] Although actual and theoretical chip feature sizes are both decreasing, the former seem to be catching up with the latter. Thus progress beyond five years could well slow down – or not. Yet, even if the sector does not pull another rabbit out of its hat, systems would still be ten times more powerful than they are today. Similar progress in the cost/value ratio of other silicon products and hard drive memory appears nearly certain.

What would greater speed alone buy? Faster language translation and graphics rendering are two possibilities (even if serious games require the use of graphics accelerators). Many computer games (including Sid Meier's *Civilization* series) employ increasingly sophisticated decision algorithms.[5] None of this, however, can yield, for example, synthetic thespians ("synthespians") or high-quality language translation without corresponding improvements in software. One might argue that once processing speed ceases to be a barrier to performance, the incentives for more sophisticated coding to wring the most out of more stubborn silicon will become clearer – but that is not guaranteed either.

Continually cheaper and more capacious storage, for its part, may make automatic and even multiple backup for information more common. This should make it nearly impossible to erase information – even if, in a crisis, it makes it more difficult to find where that backed-up information actually lies. Automatically archiving multiple copies may make it easier to roll back to an uncorrupted copy but will do little to prevent hacker-induced corruption in the first place.

With machines faster and memory cheaper, software programs are likely to bulk up to match. Unfortunately, complexity is bad for security. It creates more places for bugs to lurk, makes interactions among software components harder to understand, and increases the flow rate of packets well past where anyone can easily reconstruct what happened when things go wrong. More complex documents are likely to have much more buried

---

[4] Michael Kanellos, "New Life for Moore's Law," http://news.com.com/2009-1006_3-5672485.html, April 19, 2005.
[5] See "Mind Games" *IEEE Spectrum*, 39, 12, December 2002, pp. 40–4.

information and meta-information; the author of the Melissa virus was caught because the Microsoft Word document used to launch the virus was stamped with identifying information on who created it.[6]

Must software grow continuously more complex?[7] With Microsoft the de facto monopolist in applications software, its interpretation of the software market faces little challenge.[8] The attention being paid to Linux, at least among computer aficionados, suggests that transparency may have its fans. By contrast, Windows software hides its apparent complexity by increasing its deep complexity. So far, Linux has proven itself to be more reliable but only somewhat more secure than the Windows NT family. But few roll their own software these days and need unrestricted access to the innards of operating systems to do their job.

A.2 Cyberspace in More Places

Even were technology's leading edge to freeze in place, cyberspace would spread thanks to the declining cost of electronics, the increasing capacity of fiber optics, the clever exploitation of spectrum, and the built-in momentum of today's investments and tomorrow's plans. The number of Internet users already passed a half a billion in 2002, according to IDC, a computer consulting firm; by the end of 2006, it should have crossed a billion.

Cyberspace, at a minimum, may come to characterize media not usually associated with that term. Today, for instance, phones are phones and the Internet is the Internet. They are the two universal networks. Yet, they are becoming harder to distinguish. Phones have become digitized and the typical new cell phone comes with nontrivial LCD displays. Voice-over-IP calls are gaining market share, and with it the capability

---

[6] Stephen Shankland, "Melissa Suspect Arrested in New Jersey," http://news.com.com, April 2, 1999. The XML format of Microsoft Vista documents is a hopeful sign, though.

[7] Those who migrate, however, from mainframes and minicomputers to Linux boxes and PCs tend to travel in the direction of less complexity. Smart cell phones are simpler than PCs (but are almost always purchased in addition to rather than substitutes for them). However, the long-heralded substitution of Web access devices (or Net PCs) for general-purpose computers never happened.

[8] But see "Spot the Dinosaur," *Economist*, March 30, 2006, pp. 53–4, which argues that the company's core business is under threat from online software.

for stereo calls and more usable audioteleconferencing. Meanwhile, the circuit-switched infrastructure is mutating into a packet-switched one. How much of this convergence will be visible, apart from on phone bills? Might telephones be powerful enough to convert speech to text in real time, displacing keyboards and allowing oral requests for visual material? (For example, if you ask, "weather, please," the "phone" would return a forecast; 800 information is now mostly automated.)

Several years ago, the billionth person acquired a cell phone,[9] and chances are that person lived in a developing country. By some estimates, 2006 saw the second billionth cell phone owner. The arrival of the next few billion customers may be paced by how cheap reception devices become; until then, lagging infrastructure remains the greater constraint. Most cities with more than a million inhabitants should be on the world's fiber-optic lines a few years after Wall Street shakes off its losses from the likes of Global Crossing. Midsized cities should follow shortly thereafter.[10] This hardly implies that everyone in these cities will immediately have access to broadband services (such as video on demand), but once the costly part of laying a fiber-optic line (that is, digging the trench and so on) is finished, its economics call for putting in as many connections on it as the imagination can support. The resulting capacity will immediately and grossly exceed the capacity of any city to generate telephone calls.[11]

Telecommunications are also likely to thicken in the countryside (that is, away from fiber heads). Near-city areas may still be served via extended microwave towers, but in rural areas, communications will have to rely on outer space. Two bankruptcies, admittedly, have tempered investors' enthusiasm to finance the next Iridium or Globalstar, and thus little new capacity can be expected from low-earth orbiting satellites. Geosynchronous communications are possible if one has radar antennae pointed to a fixed spot in the sky – which means that they, too, must be fixed and

[9] Garnter, a consulting group, forecasts that by 2009, 40 percent of the world's adult population will be carrying cell phones. David Pogue, "1 Landline + 1 Cellphone = 1 Handset," *New York Times*, August 4, 2005, p, C1.

[10] Between 1996 and 2000, for instance, the cost to move a bit over fiber dropped by a factor of more than 15,000, estimated David Huber, CEO of Corvis. "At the Speed of Light," *Business Week*, October 9, 2000, p. 145.

[11] Corvis has already introduced a single-strand fiber line that promises 3 terabits per second – enough for 50 million simultaneous phone calls.

cannot move.[12] Such fixed very small aperture terminals (VSAT) feeding repeaters may be an interim path to the Internet. If, however, phased array radar comes down in price, mobile antennae may be able to tune into satellites by correcting their orientations electronically as they move.[13]

Accompanying this shift from immobile to mobile Internet connections is the insinuation of wireless connections within local area networks. Wi-Fi (or 802.11b) can deliver 11 megabits per second to base stations; Wi-Max may have five times the bandwidth.[14] The as-yet less-popular Bluetooth may enable spontaneous device-to-device networking based on certain criteria (for example, it might be instructed to alert everyone within a hundred meters who has medical skills).

This shift has several implications. First, the denser the electronics and the more that cyberspace pervades real space, the more dependent real life becomes on the correct functioning of cyberspace. Second, as early broadband experience suggests, always-on systems are always-probed ones. To recover modem-era levels of security requires that users install firewalls. Yet, many users do not and others do so with defaults left intact. Bluetooth security is almost nonexistent, with the industry relying on the weakness of signals to keep users out of trouble.[15] Third, the ability to read messages from or into networks no longer depends on the ability to hack into them or to place a wiretap on them; getting an RF device close to a network that is poorly protected may suffice. Despite the presumption

[12] Direct television broadcasters and other geosynchronous satellite owners are allocated orbital slots so that many satellites can share the same spectrum. Direct radio broadcasters such as XM and Sirius, however, have been allocated a slice of the entire spectrum (over North America) and do not require that antennae point directly at the satellite all the time – as an antenna mounted on a moving car cannot do. Many airlines (including JetBlue) have acquired antennae that permit their planes to maintain a lock on satellite signals throughout almost all of their flight.

[13] Arinc makes a tail-mounted antenna that can provide business jet passengers with broadband satellite connections; it weights only 5 kilograms. David Hughes, "The K_u Connection," *Aviation Week and Space Technology*, April 4, 2005, p. 59.

[14] In mid-2005, Sprint and BellSouth announced plans to introduce Wi-Max service, prior, however, to its standardization.

[15] At DEFCON 2004, someone demonstrated a device that could read Bluetooth devices from over a mile away (www.schneier.com/blog/archives/2005/04/bluetooth_xsnipd.html). Some thieves are using Bluetooth phones to find Bluetooth-enabled laptops in parked cars which they can then steal (www.schneier.com/blog/archives/2005/08/bluetooth_as_a.html).

that the reduction in security from land-line to over-the-air connections would automatically prompt countervailing measures (such as cryptography), cellular telephone calls remain easy to intercept and too few people seem to mind.[16] Fourth, the growing internationalization of cyberspace may, in the short term, introduce more foreign mischief makers to the medium. As more third world leaders grow conscious of how important information systems are to societies, they may give hacking a fresh look. Alternatively, the increasing importance of the Internet to third world economies may persuade their governments to pay more attention to Western concerns about security and thus cooperate on legal issues and catching the wily hacker.

A.3 Fuzzier Borders between Systems

People expect to see a strict separation between what happens within any one system (especially theirs) and what happens outside of it. This distinction not only provides some ability to isolate systems from external threats but helps in assigning blame when something goes wrong. A computer user dialing into AOL, for example, picking up e-mail, and then logging off to read it should have a good sense of the length and depth of the interaction. Certain processes, the user could reasonably believe, should not be affected by that transaction (for example, fetching e-mail would not load new software on the system or alter settings in the Windows control panel).

The distinction between inside and outside, though, is getting fuzzier by the day,[17] and not simply because of aforementioned promiscuity

---

[16] In October 2000, Intel's Jesse Walker informed the Institute of Electrical and Electronics Engineers (IEEE) that because of fundamental faults in the standard cryptographic algorithms, even key-lengths larger than the called-for 40 bits were ultimately insecure. P. J. Connolly, "Wireless Security Riddled with Flaws," *Infoworld*, June 25, 2001, p. 62.

[17] Writes Mandy Anderss:

Back in the good old days a firewall acting as your corporate gateway was more than enough perimeter protection . . . but in today's distributed environment that perimeter no longer exists. Few organizations can now delineate where their corporate networks start and stop, so the ability to protect corporate assets has become a critical task that requires rethinking the traditional security architecture. Partner access to resources such as databases and development code is one of the main reasons that the network

that wireless devices bring to networks. Some, even Microsoft, would replace one-time-purchase shrink-wrapped software with services paid for, perhaps monthly. Many Web sites already offer e-mail and calendaring services otherwise performed by packaged software. Many software programs (including virus checkers and Microsoft's Internet Explorer) periodically update themselves over the Internet. Machines such as printers and copy machines that hitherto had to be repaired on-site can be diagnosed in advance by having them make contact over the Internet. Home security systems would not work if information generated within a dwelling were not to be transmitted to others outside it.

So, where does one system stop and another begin? Why shouldn't a car, for instance, monitor its own vital signs and then forward the information to a service center that would, in turn, nag the owner that maintenance was needed. Perhaps it can adjust the car's parameters based on the latest research, air quality alerts, or fuel prices.[18] As cars go, why not bodies? Vital signs can already be continuously and remotely monitored by their doctor (or HMO). Why not go further and convert troublesome indicators into automatic parameter adjustments (for example, via embedded drug delivery systems)?[19] Monitoring need not be voluntary – some prison parolees have to wear electronic devices that report their movements when they cross specific boundaries.[20] Perhaps future parolees will be wired so that evidence of undue excitement on their part not only alerts authorities but also spurs the release of calming chemicals into the bloodstream – a model that some could easily extend to mental health patients, welfare recipients, or political dissidents.

Entrusting outsiders with system functions is hardly helpful for privacy – and privacy may be the least of all worries. Those whose bodily

perimeter is moving outward.... So before you agree to set up site-to-site connections be sure to perform a security assessment of your partner's network. Be wary of any partner that does not request the same (Mandy Anderss, *Infoworld*, February 25, 2002, p. 26).

[18] General Motor's Northstar system, for instance, permits GM to know exactly where every car so equipped is. See also "Top 10 Techno-Cool Cars," *IEEE Spectrum*, 40, 2, February 2003, pp. 30–5.

[19] For a colorful treatment, see Cathy Newman, "Dreamweavers," *National Geographic*, 203, 1, January 2003, pp. 50–73.

[20] Martha Stewart is just the most recent celebrity to be electronically tagged.

parameters can be remotely adjusted willy-nilly bet their lives on the kindness or at least probity of strangers, or worse, strange algorithms. This may be precisely why intervention monitoring may never comes to pass – or at least not without man-in-the-loop controls.

Ironically, a leading edge of the systems outsourcing parade is security wherein third parties monitor a network's vital signs and intervene if anything wrong announces itself. Thus, often poorly trained and overbooked systems administrators can be replaced with market-selected experts. Yet, many security features do not work without the active cooperation or at least compliance of users who have come face-to-face with an angry security manager one time too many.

A.4 Accepted Cryptography

The mathematics of generating virtually unbreakable public/private key pairs and unspoofable digital signatures are already well understood. Adoption in late 2001 of National Institute for Standards and Technology's (NIST) Advanced Encryption Standard has extended this difficulty to the official symmetric key standard.[21] Perhaps one day, in contradistinction to current practice, all digitized synchronous point-to-point exchanges (that is, every future phone call) will be encrypted unless specifically sent in the clear. Barring the introduction of quantum computers that can factor large prime numbers and thereby break digital signature and public key encryption methods, the fundamental parameters of cryptography are unlikely to change much over the next twenty years. Usage levels are more likely to change than will these fundamentals. In other words, the codemakers will win; the codebreakers will lose.

But how universal will this change be? Encryption (1) can be time-consuming and (2) is one more thing that can go wrong in a connection. Better hardware can fix the first problem – unless message lengths rise even faster; for example, today's text messages will become tomorrow's more bit-intensive talking heads. Good software will be required to

---

[21] It replaced the Data Encryption Standard (DES), which is considered easily breakable with today's computers. Interim improved techniques, such as triple-DES or private counterparts that had not come into universal use.

resolve the second problem. It takes a public key infrastructure to bring success to the kind of authentication required for digital signatures and for public/private key exchange – required when one or both parties need to verify themselves. It is unclear whether today's mish-mash of private infrastructures will offer enough credibility, universality, and invisibility to make digital signatures routine. Authentication may have to be a government function – just as governments are becoming more nervous about not knowing what their citizens may be saying under the cloak of anonymity. A requirement to use digital signatures to conduct certain interactions with governments (such as paying taxes) or as a replacement for notary stamps may boost the practice. Governments and other institutions could also use them to authenticate documents such as college transcripts. Some data networks distinct from but linked to the Internet may one day require that packets be signed before being transported.

Cryptographic techniques improve security and, as such, complicate information warfare significantly. Done right, they are very good security devices, albeit not perfect; for example, key management protocols are a common failure, and a compromised computer can sign false messages. Insofar as cryptography is driven by the fear of cyber-attacks, the proliferation of encryption bounds the damage that can be done, at least by amateurs.

But will information security rise? Year-to-year improvements in security depend on the delicate balance among software safeguards (notably for operating systems), new vulnerabilities introduced in place of old ones, the diligence with which black and white hats in cyberspace maintain their own alert systems for vulnerabilities, and the ease with which attack tools can be packaged and distributed. They also reflect the balance among several factors: the growing appreciation, especially among revisionist states, of what hostile conquest in cyberspace may do; the seriousness with which potential victims approach security[22]; trends in

[22] According to the *Economist*:

A survey of IT officers and CIOs carried out by Morgan Stanley after the 9/11 attacks found that security software had jumped from fifth priority or lower to become their first priority.... Chris Byrnes, an analyst at Meta group, notes that the proportion of his firm's clients (mostly large multinational corporations) with dedicated computer-security teams has risen from 20 to 40 percent in the last two years. He expects the figure

encryption; and progress in making systems fussier about what they ingest and how they digest it.

Lastly, a cautionary note: even if hard encryption coupled with the growing volume of chatter reduces the cost-effectiveness of capturing signals, the alternative of wafting or floating miniature microphones or cameras to the source of where people are talking can make up for some of what is lost by the growth of encryption.

A.5 Privatized Trust

The trustworthiness of content in and messages sent via cyberspace has been on a continuous decline ever since the medium left the cloisters of academe for the real world. Hijacked Web addresses, spam,[23] phishing, gossip at warpspeed, purloined credit cards, and repudiated credit card transactions no longer merit news coverage. DSL and cable modem owners who monitor their incoming packets report an average

to reach 60 to 70 percent within the next two years. Previously, he said, it was financial-services firms that were most serious about security but now firms in manufacturing, retailing, services are following suit ("Securing the Cloud ," *Economist*, October 24, 2002, p. 4).

But see Clay Wilson, "Computer Attack and Cyber Terrorism: Vulnerabilities and Policy Issues for Congress," *CRS Report for Congress*, October 17, 2003. Wilson cites Gary Anthes and Thomas Hoffman, "Tarnished Image," *Computerworld*, May 12, 2003, p. 37, who observed that managers in a 2003 survey pointed to the estimated $125.9 billion spent on information technology (IT) projects between 1977 and 2000 in preparation for the Y2K changeover, now viewed by some as a nonevent. Sources reported that some board-level executives stated that the Y2K problem was overblown and over funded then, and, as a result, they are now much more cautious about future spending for any new, massive IT investments.

[23] Brightmail's estimate of the percentage of e-mail considered junk more than doubled in the six months after September 2001, from 8 to 17 percent by February 2002. Stephanie Olsen, "Spam Flood Forces Companies to Take Desperate Measures," http://news.com.com/2009-1023-864815.html, March 21, 2002. By 2003, the number was up to 46 percent. Stephen Wildstrom, "Technology and You," *Business Week*, May 19, 2003, p. 20. Worried Declan McCullagh, "One of my news.com colleagues estimates that spam soon will make up the majority of message traffic on the Internet." Declan McCullagh, "Be Wary of Washington's Spam Solution," http://news.com/2010-1071-3-957024.html, September 9, 2002. By 2003, it had. The latest figures suggest that two-thirds of all e-mail is spam, a ratio that has held constant over the last two years. MessageLabs, cited in the *Economist* online edition, economist.com/agenda/displayStory.cfm?story_id=4269099, August 10. 2005.

of several hacker probes a day. By one estimate, roughly 5 to 10 percent of all Internet packets are somehow connected with planned or ongoing distributed denial-of-service attacks.[24] The efficient automation of sophisticated attack scripts, or prepackaged zombie sales, just adds to the muck.[25]

Credibility will not be helped by a proliferation of software agents – pieces of code generated by clients and used to find, request, and negotiate services such as a meals-lodging-entertainment-and-transportation package. Agents built badly, whether by error or ill will, may wreak havoc by causing servers to cycle forever for an answer, setting up mutually destructive server-to-server harmonics, or replicating without bounds. Malevolent agents may look for certain outgoing mail and then work their way into the systems from which it came. Feature interaction among agents may yield unexpected consequences. Back-propagating agents (those that enable the tracing of the source of material) could have a chilling effect on free network speech.[26] Hijacked agents or agents forwarded to successively less trustworthy servers ("a friend of a friend of a friend . . .") may relay misleading or dangerous information back. Replicating agents may carry replicating viruses whose eradication may be fiendishly difficult if there are niches in the network ecosystem that cannot be identified or do not cooperate.

The rise of instant messaging, affinity group Web sites (such as those that serve people suffering from the same rare disease), blogs, and listservs (e-mail lists) portend at least a partial shift back from public to private sources of news. Originally most people learned most of what they knew of the world from those they trusted personally. With the rise of literacy came mass media, now on continuous loop thanks to radio and television stations, whose news-streams are, in turn, reflected in cyberspace. Person-to-person messaging is becoming not only comparably efficient

---

[24] Robert Lemos, "A Year Later, DDOS Attacks Still a Major Web Threat," http://news.com/news/0-1003-201-4735597-0.html, February 7, 2001.

[25] Some hackers make money by converting other people's computers to zombies and selling access to thousands of them at a time to others interested in launching distributed denial-of-service attacks. See, for instance, Stephen Labaton, "An Army of Soulless 1s and 0s," http://news.com.com/2100-7349_3-5761553.html, June 24, 2005; Joris Evers, "Hacking for Dollars," http://news.com.com/2100-7349_3-5772238.html, July 6, 2005.

[26] There is already evidence of this potential in the record industry's attempts to have Congress legitimize mischief against music swappers.

at broadcasting, but also easier to tailor and inherently more trustworthy – especially for the kind of news "you can use," such as news on health, neighborhood conditions, education, transportation, and commercial opportunities.

The dirtier cyberspace is, the less people, apart from the most credulous, are apt to trust the information it holds. They will react gingerly to what they receive, and turn to sources with which they are familiar. The search for redundancy and trust would make it harder for a remote one-time information attack to sway a large body of opinion. Indeed, the very fractionation of a nation's body politic into linked clusters complicates any outsider's ability to manipulate it using mass media techniques.

A.6 The Possible Substitution of Artificial for Natural Intelligence

Artificial intelligence is a typical half-empty/half-full story. Predictions that computer-based logic could replace human logic soon have given new meanings to the word "soon." But progress has been made. Computer prowess at playing chess, simulating experts in making well-ordered decisions, converting English speech into text, understanding domain-limited inquiries, and translating among languages has waxed year by year although never enough to make humans completely redundant, perhaps chess aside. Further improvement may make today's barriers to international commerce such as differences in languages (at least if limited to business conversation) or monetary currencies disappear. Nevertheless, the ability to reproduce cultural or contextual nuances with sufficient fidelity will remain elusive; the same holds for pattern recognition.

One oft-sought goal, in a world crowded with information, is the ability to evaluate, categorize, and forward text to those with an expressed interest in a topic. Today's methods are primitive; tomorrow's may be slightly better, but not necessarily good enough to replace human editors – although some of what Google.com and its clones do is fairly sophisticated. A similar capability may be to sort through e-mail to filter the important ones up for immediate attention and filter out the spam.[27]

---

[27] This approach is called whitelisting. See "The 2nd Annual Review of Ideas," *New York Times Magazine*, December 15, 2002, p. 136.

The cleverness of natural intelligence in getting the spam through raises the bar that artificial intelligence (AI) must hurdle.[28]

Researchers, notably the late Michael Dertouzos of the Massachusetts Institute of Technology (MIT),[29] have worked hard to make computers as simple to use and as intuitive as butlers: a user can express a general intention with sufficiently clarity, and the computer does what the user needs done (for example, making complex travel arrangements). Unfortunately, most attempts to make computers friendlier end up making them more intuitive for people with common needs and less yielding for those who want to do things slightly differently. Good grammar checkers are still a dream; they simply cannot determine whether a word has been used correctly in its context.

The most ambitious system dream is that of a general knowledge engine. Progress here, more than anything else, would bring cyberspace into its own as a realm where semantics are employed in reaching conclusions. Fed facts (such as those scraped off the Web), the engine would deduce their meaning and reach conclusions or make predictions based on their analysis. When coupled with a user's preference matrix, such an engine could suggest decisions.[30] DoD, for its part, may want an intelligence fusion engine that can feed a course-of-actions model.

Here, too, the ability to displace humans is likely to be proportional to the narrowness of the domain under consideration and the precision to which inputs can be specified. Computers can generate a route

---

[28] At least as of early 2006, the spam filters associated with Microsoft's Hotmail seemed to be ahead.

[29] He was no doubt thinking of the work he and his team had done on his Project Oxygen, funded by DARPA.

[30] This depends, of course, on whether these Web sites have not been tampered with. Reports *Business Week*:

> CIA staffers have been caught altering entries on Wikipedia. . . . Someone using an agency computer changed Wiki's Clinton entry to note that the ex-President was "dumber" than his GOP predecessors. Spooks aren't the only ones playing dirty tricks. Wiki reports that computer users at the Justice Dept., Marine Corps, and Navy have politicized entries in recent weeks. Earlier this year, Wikipedia blocked Capitol Hill access to the site after lawmaker entries were subjected to political spin and fabrications. Wikipedia founder Jimmy Wales calls the shenanigans, "routine" (*Business Week*, March 13, 2006, p. 49).

based on geography. If fed traffic patterns and driver preferences (for example, being assuredly ten minutes late beats a 10 percent chance of being thirty minutes late), they may one day outperform people. But then throw in apparently extraneous considerations, such as the opportunity to see a great sunset from a specific vantage point, the disinclination to encounter someone unpleasant, or a last-minute reminder to stop at a store. Satisfying the sort of considerations that people weigh every day would require considerable sophistication on the part of software writers.

Computers are presently literal devices that do what they are told without many questions (apart from the occasional "are you sure?" when users try to make permanent changes such as erasing files). Someday they may work differently, especially if placed in tricky and deceiving environments. By basing their decisions not so much on commands but on a flow of facts against a background of criteria, they could ensure the results of their actions stay within specific parameters. They might go further to test whether commands or assertions of fact are anomalous; if so, they could ask for more authentication or evidence to back them up (just as credit card companies already do for some purchases). Alternatively, commands could be accompanied by rationales that would have to be checked out before such commands are carried out. Such a model entails sophistication, and it is unclear whether building a spoof-resistant inference engine is more cost-effective than implementing a reliable security model from the beginning.

Learning systems generate rules from facts. Neural nets build association matrices tuned by backward propagation from examples. Both may permit computers to get smarter with every new experience. Yet, if the experience on which their progress is based is corrupted, then corrupted conclusions may follow. Finding and fixing corruption will be hard; detecting corruption may be possible only by observing poor decisions, and such decisions may be triggered only by rarely encountered inputs. Even if corruption is found, one must determine when the system was corrupted, so as to return it to its last known uncorrupted state, if one has archived such parameters. Otherwise, the system has to be returned to its original settings, which means losing all the learning that has taken place since it was turned on.

A.7 Conclusions

Four primary lessons of relevance can be drawn from this sketch of the undiscovered country to come.

First, as subsections A.1 and A.2, as well as sections 6.4.1 and 8.2, suggest, cyberspace will continue to grow bigger, more complex, and more pervasive. There will be a good deal more of it, and chances are that more and more human functions will be conducted in and through it. The world will grow continually more dependent on their proper functioning. Information-free zones (that is, private spaces) will shrink.

Second, the very definition of an enterprise infrastructure will continue to spread – internally through the consolidation and integration of underlying functions, and externally to business partners and customers. Formerly hard distinctions between the systems one has control over and the rest of the world will soften, whether because of unrestricted RF communications or through deliberate migration of functions from packaged software to network services. The fundamental vulnerability to hostile action will rise – but so will the ability to express an organization's culture through the architecture of its information infrastructure.

Third, a comparison of subsections A.4 and A.5 suggests a potential divergence between private and public versions of cyberspace. If and when cryptography (such as for private virtual networks) and digital signatures take hold, enterprises should realize a degree of protection from the turbulence of cyberspace writ large. But, while the trustworthiness of information in public cyberspace may continue to deteriorate, people may find private ways to navigate safely in such waters.

Fourth, and most speculatively, if machines really can think intelligently (pace A.6) they will be given responsibilities that humans alone now have. Perhaps sooner than that, they will stand between the user and cyberspace, filtering out the false and worthless and sorting the rest out for human consumption. Perhaps later, they will be able to think about what they receive, using rules designed to manipulate semantics to categorize information and perhaps even draw conclusions from it.

So, in fits and starts, cyberspace is acquiring consequence – and as it does, so does the struggle to seek conquest in it.

Index

accessing cyberspace, 294
accessing systems for information
    warfare, 75–9
ActiveX, 242, 287
Acxiom, 202
Adams, James, 38
affinity cards, 196, 209
Afghanistan war
    al Qaeda's dependence on Internet
      and, 45–6
    coalitions and, 129, 172
    commercial satellite photos and,
      177
    information overload and
      decision-making hierarchy,
      109
    minimal role of cyberspace
      conquest in, 2
agoras. *See* castles and agoras
Ahmad, Fahim, 261
AI. *See* artificial intelligence
air bags, 197
air-gapped systems, 64, 106, 270
al Qaeda. *See also* specific operations,
    e.g, September 11, 2001,
    attacks
    command and control exercised
      online, 48
    dependence on Internet,
      45–6

relative dependence on
    information infrastructure,
    272
Alexander, Yonah, 44
alliances. *See* coalitions
Alneda.com, 259
Amazon.com, 146
Amerindian language, Sapir-Whorf
    hypothesis based on, 235–6,
    248
Andalusi, Ameer, 261
Anna Kournikova virus, 92, 241
Annan, Kofi, 98
ANSI X12 standard for electronic
    data interchange (EDI), 246
anticorruption uprisings in Southeast
    Asia, 148
architecture
    boundaries between systems,
      growing fuzziness of, 306
    defined, 142–3
    enterprise architecture and systems
      integration, 142–8
    form and function, relationship
      between, 240–1
    informational. *See* information
      architecture
    standards defining, 244
Arthur, Brian, 277
artificial intelligence (AI), 115, 248

Silicon Valley, 64, 128, 156
simple object access protocol (SOAP),
    254
SIPRnet, 64
SOAP. *See* simple object access
    protocol
social norms operating in cyberspace,
    8
soft power vs. hard power, 3–4, 276
software and hardware
    powerful hardware leading to
        complex software, 292–4
    redundancy, 84–5
Sony, 277
spam, 15, 48, 73, 112, 115, 213, 274
specialists, distillations of
        information from, 113
speed of processors, 292–3
Spot satellite, 175
spying or eavesdropping
    assessment as countermeasure
        against, 88–9
    information overload, effects of,
        105–6
    intimacy and vulnerability to, 229
    policy on use of hostile conquest
        and, 260–1
    strategic means of information
        warfare, 79–80
    type of information warfare,
        28–9
    wholesale level, operation at, 193
standards
    architecture of cyberspace defined
        by, 244
    coalition dependency and, 137–8,
        140–1
    encryption, 299–301
    geospatial database, 174–5
    national identity systems, 190–1
    policy questions regarding friendly
        conquest and, 277–8, 281
    semantic layer, 245–9
    XML, 245–7, 254
Stasi, 209

stealth technology development,
    100
Stephenson, Neil, 5
storage and use, safety issues relative
    to, 99–100
strategic information warfare,
    73–101
Sun Microsystems, 254, 285
Sun Tzu, 38
Surrey Satellite Technology, 175
surveillance devices, proliferation of,
    196–8
syntactic layer, 8–9
    boundaries with semantic layer
        friendly conquest, 242–4
        hostile conquest, 241–2
        problems maintaining, 240
    cyberspace, 236–7
    human language, 232
    information warfare, 24–5
    interoperability of layers and,
        237–8
    OSI model, 239
    XML, 245–7, 254
systems integration and enterprise
    architecture, 142–8

Tanweer, Shehzad, 47
Teapot Dome scandal, 230
technological development, ecology
    of, 155–9
telephony. *See* cell phones
temporary vs. permanent damage
    from computer network
    attacks, 37
TerraSAR-X satellite, 176
terrorists and terrorism
    hostile conquest used against,
        259–62
    information warfare and, 43–6
    layers of cyberspace used by Islamic
        terrorists, 10
    narrative, role of, 216
    national identity systems and, 169,
        183, 189–90